# SPRINGER LABORATORY

**Springer Laboratory Manuals in Polymer Science**

Pasch, Trathnigg: HPLC of Polymers
ISBN: 3-540-61689-6 (hardcover)
ISBN: 3-540-65551-4 (softcover)

Mori, Barth: Size Exclusion Chromatography
ISBN: 3-540-65635-9

Pasch, Schrepp: MALDI-TOF Mass Spectrometry of Synthetic Polymers
ISBN: 3-540-44259-6

Kulicke, Clasen: Viscosimetry of Polymers and Polyelectrolytes
ISBN: 3-540-40760-X

Hatada, Kitayama: NMR Spectroscopy of Polymers
ISBN: 3-540-40220-9

Brummer, R.: Rheology Essentials of Cosmetics and Food Emulsions
ISBN: 3-540-25553-2

Mächtle, W., Börger, L.: Analytical Ultracentrifugation of Polymers
and Nanoparticles
ISBN: 3-540-23432-2

Heinze, T., Liebert, T., Koschella, A.: Esterification of Polysaccharides
ISBN: 3-540-32103-9

Koetz, J., Kosmella, S.: Polyelectrolytes and Nanoparticles
ISBN: 3-540-46381-X

Stribeck, N.: X-Ray Scattering of Soft Matter
ISBN: 3-540-46488-4

Norbert Stribeck

# X-Ray Scattering
# of Soft Matter

With 92 Figures and 6 Tables

 Springer

Norbert Stribeck

Universität Hamburg
Institut für Technische und
Makromolekulare Chemie
Bundesstr. 45
20146 Hamburg
Germany
*e-mail: norbert@stribeck.de*

Library of Congress Control Number: 2007922403

DOI 10.1007/978-3-540-69856-2

ISBN 978-3-540-69855-5 Springer Berlin Heidelberg New York

e-ISBN 978-3-540-69856-2

Springer is a part of Springer Science+Business Media
springer.com

© Springer-Verlag Berlin Heidelberg 2007

Typesetting: Camera-ready copy from the author
Data conversion and production by LE-TEX Jelonek, Schmidt & Vöckler GbR, Leipzig, Germany
Cover design: eStudio Calamar, Girona/Spain

SPIN 11533863     55/3100/YL - 5 4 3 2 1 0     Printed on acid-free paper

# Springer Laboratory Manuals in Polymer Science

**Editor**

Priv.-Doz. Dr. Harald Pasch
Deutsches Kunststoff-Institut
Abt. Analytik
Schloßgartenstr. 6
64289 Darmstadt
Germany
e-mail: hpasch@dki.tu-darmstadt.de

**Editorial Board**

PD Dr. Ingo Alig
Deutsches Kunststoff-Institut
Abt. Physik
Schloßgartenstr. 6
64289 Darmstadt
Germany
email: ialig@dki.tu-darmstadt.de

Prof. Josef Janca
Université de La Rochelle
Pole Sciences et Technologie
Avenue Michel Crépeau
17042 La Rochelle Cedex 01
France
email: jjanca@univ-lr.fr

Prof. W.-M. Kulicke
Inst. f. Technische u. Makromol. Chemie
Universität Hamburg
Bundesstr. 45
20146 Hamburg
Germany
email: kulicke@chemie.uni-hamburg.de

# Preface

Mehr Licht!

*(J. W. v. Goethe)*

The application of X-ray scattering for the study of soft matter has a long tradition. By shining X-rays on a piece of material, representative structure information is collected in a scattering pattern. Moreover, during the last three decades X-ray scattering has gained new attractivity, for it developed from a static to a dynamic method.

The progress achieved is closely linked to the development of both powerful detectors and brilliant X-ray sources (synchrotron radiation, rotating anode). Such point-focus equipment has replaced older slit-focus equipment (Kratky camera, Rigaku-Denki camera) in many laboratories, and the next step of instrumental progress is already discernible. With the "X-ray free electron laser" (XFEL) it will become possible to study very fast processes like the structure relaxation of elastomers after the removal of mechanical load.

Today, structure evolution can be tracked *in-situ* with a cycle time of less than a second. Moreover, if a polymer part is scanned by the X-ray beam of a microbeam setup, the variation of structure and orientation can be documented with a spatial resolution of 1 $\mu$m. For the application of X-rays no special sample preparation is required, and as the beam may travel through air for at least several centimeters, manufacturing or ageing machinery can be integrated in the beamline with ease.

On the other hand, the result of the scattering method is not a common image of the structure. There is not even a way to reconstruct it from scattering data, except for the cases in which either anomalous scattering is employed, or a diffraction diagram of an almost perfect lattice structure is recorded. Because most of the man-made polymer materials suffer from polydispersity and heterogeneity, the crystallographic algorithms of structure inversion are in general restricted to the field of biopolymers (e.g., protein crystallography). Thus the ordinary polymer scientist will deal with scattering data rather than with diffraction data. These data must be interpreted or analyzed. This book is intended both to guide the beginner in this field, and to present a collection of strategies for the analysis of scattering data gathered with modern equipment. Common misunderstandings are discussed. Instead, advanced strategies are advertised.

An advantage of a laboratory-oriented textbook is the fact that many technical aspects of our trade can be communicated[1]. Their consideration may help to improve the quality and to assure the completeness of the recorded data. On the other

---

[1] An example is the chapter entitled "It's Beamtime, Phil". It is written in the hope that in particular the practical work of students will benefit from it.

hand, the concept is restricting the presentation of the mathematical background to a terse treatment. For a field like the scattering that is virtually interpenetrated by mathematical concepts this is not unproblematic. As a consequence, it was impossible to present mathematical deductions, which could have been an assistance to methodical development by the reader. In this respect even the references given to original papers are not really helpful, because in such publications the fundamental mathematical tools are expected to be known. Nevertheless, this restriction may be advantageous from a different perspective. The terse scheme is enhancing the presentation of the fundamental ideas and their repetitive use in different subareas of the scattering technique.

This book with its special focus on application was stimulated by a suggestion of Prof. Dr.-Ing. W.-M. Kulicke. I greatly appreciate his support. Moreover, the manuscript has its roots in thirty years of practical work in the field of scattering from soft materials conducted in several labs and at several synchrotron sources. During this time the author has assisted many external groups with their practical work at the soft-matter beamlines of the Hamburg Synchrotron Radiation Laboratory (HASYLAB at DESY), supported evaluation of scattering data, and worked as a referee in the soft-condensed matter review-committee of the European Synchrotron Radiation Facility (ESRF) in Grenoble. The accumulated handouts prepared during twenty years of lecturing scattering methods at the University of Hamburg have been a valuable source for the book manuscript.

There are many other people who have – in different respect – contributed to this work. The first to mention is my teacher, Prof. Dr. W. Ruland. I am grateful for his art of teaching the scattering. Wherever in this book I should have been able to explain something clearly and concisely, it is his merit. The second to mention is Prof. Dr. H. G. Zachmann. In his group I enjoyed to become involved in many practical issues of soft matter physics. In particular I appreciate many helpful comments on the manuscript that have been supplied by Prof. Dr. W. Ruland, Dr. C. Burger, Prof. Dr. A. Thünemann and Prof. Dr. S. Murthy. In addition, there are many other colleagues who have stimulated my work by fruitful cooperation, discussion and support. To mention them all would fill pages.

The complex task of writing a scientific manuscript has been significantly eased by authoring tools that keep track of the formal aspects of the growing manuscript. For this reason I thank the developers of LyX, Koma-Script and LaTeX (in particular Matthias Ettrich and Markus Kohm) for their free and superb software. Moreover, I highly appreciate the excellent guidance and the distinguished manuscript editing by the team at Springer Publishers.

Last but not least I express cordial thanks to my wife Marie-Luise and to my children for their continuous support.

Hamburg, January 2007                                                       N. Stribeck

# List of Symbols and Abbreviations

The handling of polar coordinates is a general problem in a book on scattering, where the symbol $\theta$ that is normally used to indicate the polar angle is already used to indicate the Bragg angle. Too late I became aware of the problem and tried to introduce a consistent notation. Unfortunately the problem was more involved than I thought, as colleagues pointed out after proofreading the manuscript. Based on suggestions I finally tried to harmonize the nomenclature. Nevertheless, the reader should be aware of possible remnant inconsistencies concerning the use of the symbols $\psi$, $\varphi$ and symbols of related angles.

| | |
|---|---|
| $\langle\rangle$ | Averaging operator |
| $\langle\rangle_V$ | Irradiated volume average |
| $\langle\rangle_\omega$ | Solid-angle average |
| $\sqcap$ | Slice mapping |
| $\{\}$ | Projection mapping |
| $\star$ | Convolution operator |
| $*\varphi$ | Angular convolution |
| $\otimes$ | Correlation operator |
| $\star 2$ | Autocorrelation operator |
| $*$ | Complex conjugate. $z = a + ib$; $z^* = a - ib$ |
| $\nabla$ | Gradient operator |
| 1D | One-dimensional |
| 2D | Two-dimensional |
| 3D | Three-dimensional |
| $A(\mathbf{s})$ | Scattering amplitude |
| $\mathbf{a}$ | Scaling vector (anisotropic dilation) |
| $a$ | Scaling factor (isotropic or 1D dilation) |

| | |
|---|---|
| $a$ | In a lattice: edge length of unit cell, i.e., the distance between the $\delta$ ( )-elements that make the abstract lattice $c$ ( ) |
| $\alpha_i$ | Angle of incidence on the sample surface |
| $\alpha_e$ | Angle of exit from the sample surface |
| $B(h)$ | Integral breadth of the distribution $h$ |
| $c$ ( ) | Comb function (abstract lattice) |
| CLD | Chord length distribution $g(r) = -\ell_p\,\gamma''(r)$ |
| CCD | Charge-coupled device |
| CDF | Chord distribution function $z(\mathbf{r}) \propto -\Delta\gamma(\mathbf{r})$ |
| $\delta$ ( ) | DIRAC's delta function |
| $\Delta$ | Laplacian operator |
| DESY | **D**eutsches **E**lektronen-**SY**nchrotron (Hamburg, Germany) |
| DI | Digital image processing |
| $D$ | Fractal dimension |
| $d_{hkl}$ | Lattice repeat in WAXS (distance between net planes of a crystal indexed by $hkl$) |
| DDF | Distance distribution function |
| ESRF | **E**uropean **S**ynchrotron **R**adiation **F**acility (Grenoble, France) |
| $\varepsilon$ | Mechanical elongation ($\varepsilon = l/l_0 - 1$) |
| $\exp(-\mu\ell)$ | Linear absorption factor |
| $\mathscr{F}(\mathbf{s})$ | Fourier transform |
| $\mathscr{F}_n$ ( ) | $n$-dimensional Fourier transform |
| $\mathscr{F}_{-n}$ ( ) | $n$-dimensional Fourier back-transform |
| $f_P$ | Polarization factor |
| $f_{or}$ | Uniaxial orientation parameter (HERMANS' orientation function) |
| *FIT2D* | Scattering data evaluation program by A. Hammersley (ESRF) |
| FLASH | Free Electron Laser Hamburg |
| FWHM | Full width at half-maximum |

$g(r)$                (Radial) chord length distribution (CLD)

$g_1(x)$           (One-dimensional) interface distribution function (IDF)

GEL            Image data format returned by image plate scanners

$\gamma(\mathbf{r}) = \rho^{*2}(\mathbf{r})/k$  Normalized correlation function

HASYLAB      Hamburg Synchrotron Radiation Laboratory

$h()$            Some kind of distribution function

$hkl$             MILLER's index of a crystal reflection in reciprocal space

$(h)$             Order of a reflection, line or peak. Short for $hkl$

$H()$            Fourier transform of the distribution $h()$

$h_H(a)$          Size distribution (of particles, clusters)

$\Im()$            Imaginary part of a complex number

$I(\mathbf{s}) = \mathscr{F}_3\left(\rho^{*2}(\mathbf{r})\right)$  Scattering intensity

$I_0$             Incident intensity (i.e. primary beam intensity)

$I_t$             Transmitted intensity behind the sample

IDL            Commercial programming system for image data processing

ImageJ        Open-source programming system for image data processing

$J(s_3) = \lceil\{I\}_2(s_2,s_3)\rceil_1(s_3)$  Slit-smeared scattering intensity

$J_i$             Bessel function of the first kind and order $i$

$k = \int I(\mathbf{s})\,d^3s = \rho^{*2}(0)$  Scattering power

$L$              Lattice repeat (in SAXS: long period, in WAXS identical to $d_{hkl}$ according to Bragg's law)

$\ell$              Path of the photon through the sample

$\ell_p$           Chord length related to size of crystals or domains

$l$              In straining experiments: actual length of the sample

$l_0$             In straining experiments: initial length of the sample

$\lambda$              X-ray wavelength

$\lambda_d$           Draw ratio $\lambda_d = l/l_0 = \varepsilon + 1$

Linac         Linear accelerator

| | |
|---|---|
| $M$ | Molecular mass |
| MAXS | Middle-angle X-ray scattering |
| $\mathscr{M}()$ | Mellin transform |
| $\mu$ | Linear absorption coefficient |
| $\mu_i$ | $i$-th central moment of a distribution function |
| $\mu_i'$ | $i$-th moment about origin of a distribution function |
| OTOKO | Scattering curve evaluation program by M. Koch (EMBL, Hamburg) |
| $pv$-$wave$ | Commercial programming system for image data processing |
| $P(\mathbf{r}) = \rho^{*2}(\mathbf{r})$ | Patterson function |
| $p(r)$ | (Radial) distance distribution function $p(r) = r^2\gamma(r)$ |
| $\Phi(\mathbf{s})$ | Fourier transform of a shape function $\Phi(\mathbf{s}) = \mathscr{F}(Y(\mathbf{r}))$ |
| $\mathbf{q} = 2\pi\mathbf{s}$ | Alternate scattering vector |
| $Q = k/V$ | Invariant (SAXS) |
| $Q_P$ | Polarization quality (of a synchrotron source) |
| $\mathbb{R}$ | The set of real numbers |
| $\mathbb{R}^n$ | The $n$-dimensional vector space |
| $\mathfrak{R}()$ | Real part of a complex number |
| $R$ | Sample-to-detector distance |
| $R_g$ | Guinier radius (i.e. radius of gyration) |
| $\mathbf{r} = (r_1, r_2, r_3)$ | Real space vector |
| $r_e$ | COMPTON's classical electron radius ($2.818 \times 10^{-15}$m) |
| ROI | Region of interest (from Digital Image Processing) |
| $\rho_m$ | Mass density |
| $\rho(\mathbf{r})$ | Electron density (in the field of SAXS: deviation of the electron density from the average electron density) |
| $\rho^{*2}(\mathbf{r}) = k\gamma(\mathbf{r})$ | (SAXS) correlation function |
| $\langle\rho\rangle_V$ | Average electron density |

| | |
|---|---|
| $s$ | Magnitude of the scattering vector |
| $\mathbf{s} = (s_1, s_2, s_3)$ | Scattering vector in Cartesian coordinates |
| $\mathbf{s} = (s, \phi, \psi)$ | Scattering vector in polar coordinates ($\phi$ polar angle, $\psi$ azimuthal angle). – See the preamble to this "List of Abbreviations" |
| SAXS | Small-angle X-ray scattering |
| S/N | Signal-to-noise ratio |
| SSRL | Stanford Synchrotron Radiation Laboratory |
| $\sigma$ | Standard deviation |
| $\sigma^2$ | Variance |
| $t$ | Sample thickness |
| $t_{opt}$ | Optimum sample thickness |
| TIFF | Tagged Image File Format |
| TOPAS | Scattering curve evaluation program by N. Stribeck |
| $\theta$ | Bragg angle (half of the scattering angle) |
| $2\theta$ | Scattering angle |
| $\theta_c$ | Critical angle of total reflection |
| USAXS | Ultra small-angle X-ray scattering |
| USB | Universal Serial Bus (an interface to couple external devices to computers) |
| $V$ | The sample volume irradiated by the X-ray beam |
| VFC | Voltage-to-frequency converter |
| VUV | Vacuum ultra-violet light |
| $W$ | Beam cross-section of the incident X-ray beam |
| $x$ | Principal axis of uniaxial structure, depth in which a photon is scattered |
| XFEL | X-ray free electron laser |
| $Y(\mathbf{r})$ | Shape function ($Y(\mathbf{r}) = 0$ outside the body, $Y(\mathbf{r}) = 1$ inside) |
| $Y_H(x)$ | Heaviside function. $Y_H(x > 0) = 1$, $Y_H(x < 0) = 0$. $\partial Y_H(x)/\partial x = \delta(x)$ |
| WAXS | Wide-angle X-ray scattering |
| $z(\mathbf{r}) = -\Delta P(\mathbf{r})$ | Chord distribution function |

# Table of Contents

# 1 Polydispersity and Heterogeneity

The heterogeneity immanent to materials that show scattering but not diffraction patterns should not be ignored. An assessment concerning the significance of results can only be expected if the collected data are complete (cf. Sect. 8.4.2) and show low noise (exposure time long enough). Whenever a measured parameter value is discussed, heterogeneity results in fundamental questions to be answered: What kind of average does my method return? Is it possible to determine the width and skewness of the parameter value distribution? A brief review of such "probability distributions" and their moments is given for later reference.

## 1.1 Scattering, Polydispersity and Materials Properties

Except for biopolymers, most polymer materials are polydisperse and heterogeneous. This is already the case for the length distribution of the chain molecules (molecular mass distribution). It is continued in the polydispersity of crystalline domains (crystal size distribution), and in the heterogeneity of structural entities made from such domains (lamellar stacks, microfibrils). Although this fact is known for long time, its implications on the interpretation and analysis of scattering data are, in general, not adequately considered.

DEBYE & MENKE (1931) [1]: "It is futile to draw distinct conclusions if genuine scattering curves are not at hand. It is insufficient under any circumstances if authors state that an interference maximum or several of them exist at certain angular positions. Only a continuous scattering pattern can be the fundament of proper reasoning. Concerning the abundant reports on disordered materials it must unfortunately be stated that they are unsatisfactory in this respect. Although even in this way, by mere accumulation of data and comparison of

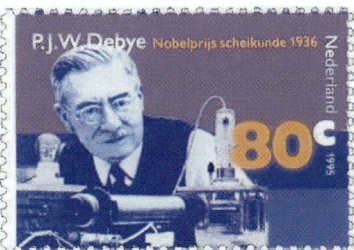

**Figure 1.1.** P. Debye (1884-1966) and his small-angle light-scattering device on a Dutch stamp

data from materials with similar chemical composition, some valuable conclusion was drawn with a higher or a lower level of significance. This situation is the result of the fact that we are insufficiently informed on the theory of the arrangement of molecules in a fluid. Only if it were possible to theoretically describe this arrangement in a similar manner as can be done for the arrangement of atoms in a crystal, it would be sufficient to report interference maxima."

**Heterogeneity.** In reality, structure is frequently heterogeneous. For example, if colloidal crystals have been produced by means of nanotechnology, it must be assumed that the material is not perfect. Thus it is of some importance to describe the deviation of the individual sample from the ideal material. For such purposes scattering methods are frequently employed and the scattering patterns are *qualitatively* interpreted. Nevertheless, the mechanisms of structure formation remain obscured as long as the amount of heterogeneity cannot be determined *quantitatively* during the structure formation process.

Different kinds of heterogeneity can be imagined. In the most simple case only a few differing structural entities are found to coexist without correlation inside the volume irradiated by the primary beam. In this case it is the task of the scientist to identify, to separate and to quantify the components of such a multimodal structure. In an extreme case heterogeneity may even result in a fractal structure that can no longer be analyzed by the classical methods of materials science.

**Polydispersity.** Quite frequently many *different but similar* structural entities can be found in a material. This is the common notion of polydispersity. Thus polydispersity means that every structural unit in the sample can be generated by compression or expansion (dilation) from a template. This building principle is mathematically governed by the Mellin convolution [2], which generates the observed structure from the template structure and its size distribution. The determination of the latter is a major goal in the field of materials science. Considering the simple case of pure particle scattering, the searched size distribution is the particle dimension distribution [3]. If, for example, the studied particles are spheres, the number distribution of sphere diameters would be of interest, and the material would advantageously be characterized by the mean diameter and the variance of the sphere diameters. Moreover, even a value describing the skewness of the sphere diameter distribution may become important in order to understand property variations of different materials.

## 1.2 Distribution Functions and Physical Parameters

A general principle is governing the relation between physical parameters and underlying distribution functions. Its paramount importance in the field of soft condensed matter originates from the importance of polydispersity in this field. Let us recall the principle by resorting to a very basic example: molecular mass distributions of polymers and the related characteristic parameters.

### 1.2.1 The Number Molecular Mass Distribution

In the basic molecular mass distribution, $N(M)$, the number $N$ of molecules in a sample is plotted *vs.* their molecular mass, $M$. Figure 1.2 presents a sketch of a

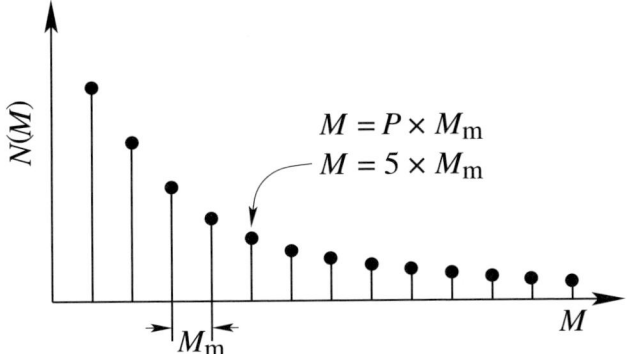

**Figure 1.2.** A number molecular mass distribution $N(M)$ of an ideal chain polymer. $N(M)$ is defined for integer multiples of $M_m$, the monomer mass. The integer factor, $P$, is called the degree of polymerization

molecular mass distribution. For ideal chains the distribution is a discrete function which is only defined for integer multiples of the monomer mass, $M_m$. The function is called the *number* molecular mass distribution, because it exhibits the number of molecules with a certain molecular weight $M$.

The function $N(M)$ can be considered a continuous function, if the average molecular weight of the chains is high enough. In this case we draw a continuous line through the points in Fig. 1.2.

It is reasonable to normalize $N(M)$ with respect to the total number of molecules in the sample

$$n(M) = N(M) \Big/ \int_0^\infty N(M) \, dM. \tag{1.1}$$

Now the function displays the number fraction of molecules with a certain molecular mass. Its integral is 1 by definition. Nevertheless, we still call it the number molecular weight distribution because the factor $\int N(M) \, dM$ is nothing but a constant.

### 1.2.2 The Number Average Molecular Mass

The obvious definition of the number average, $M_n$, of the distribution is the position on the $M$-axis that divides the area under the $n(M)$– curve in equal parts (cf. Fig. 1.3). Because of the fact that $n(M)$ is normalized to 1, each of the subareas is equal to 0.5. As 50% of all the molecules are shorter than $M_n$, the other 50% are longer than $M_n$. Bearing in mind the normalization, the number average molecular mass is

$$M_n = \int_0^\infty M n(M) \, dM. \tag{1.2}$$

This equation is, as well, the definition of the mean (cf. ABRAMOWITZ [4] chap. 26) – the first moment of the distribution $n(M)$ about origin. In fact, with respect to a

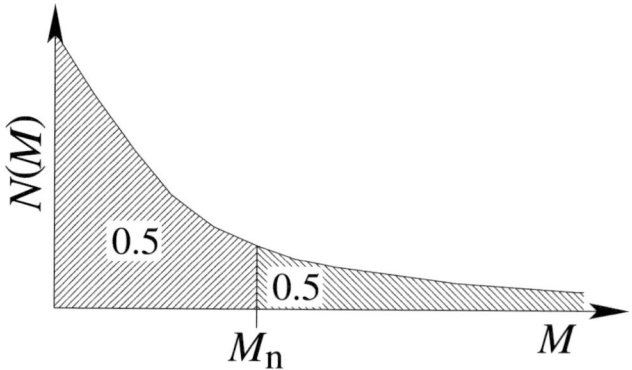

**Figure 1.3.** The number average molecular mass, $M_n$, is the position that divides the area under the corresponding distribution in equal parts

normalized distribution ($\int n(M)\,dM = 1$) the mean is the *center of gravity* of the distribution.

In order to describe the discussed distribution function, three characteristic parameters are used in polymer science. They are named number average[1], weight average ($M_w$), and centrifuge average ($M_z$)

$$M_n = \frac{\int M^1 n(M)\,dM}{\int M^0 n(M)\,dM} \tag{1.3}$$

$$M_w = \frac{\int M^2 n(M)\,dM}{\int M^1 n(M)\,dM} \tag{1.4}$$

$$M_z = \frac{\int M^3 n(M)\,dM}{\int M^2 n(M)\,dM} \tag{1.5}$$

This series of equations demonstrates a general principle in physics, namely how measurable materials parameters are generated from moments of the related distribution function.

## 1.3 Moments

The $i$-th *moment* (about origin) of a distribution $h(x)$ is defined by

$$\mu_i'(h) = \int x^i h(x)\,dx \tag{1.6}$$

(ABRAMOWITZ [4] chap. 26). We have demonstrated that the structure parameters of a polydisperse structure are closely related to these moments. $\mu_0'(h)$ is the norm

---

[1]This is the center of gravity of the distribution $n(M)$.

and $m(h) := \mu'_1(h)$ the mean of the distribution on which the definition of *central moments*

$$\mu_i(h) := \int (x - m(h))^i \, dx \qquad (1.7)$$

is based. As a measure of distribution width it is common to report the variance

$$\sigma^2(h) := \mu_2(h) \qquad (1.8)$$

or the standard deviation, $\sigma(h)$. $\mu_3(h) / \sigma^3(h)$ is known as skewness of the distribution (ABRAMOWITZ [4] chap. 26).

**Application in the Field of Scattering.** Let us consider two important distribution functions, $h_c(x)$ and $h_L(x)$. These functions shall describe the thicknesses of crystalline layers and the distances (long periods) between them, respectively. In this case we take into account polydispersity of the crystalline layers, if (at least) the two parameters $\bar{d}_c$ and $\sigma_c / \bar{d}_c$ are determined which are defined as the average thickness of the crystalline layers,

$$\bar{d}_c = \frac{\mu'_1(h_c)}{\mu'_0(h_c)},$$

and the relative standard deviation of the crystalline layer distribution,

$$\frac{\sigma_c}{\bar{d}_c} = \frac{\sigma(h_c)}{\bar{d}_c}.$$

In the classical treatment of the paracrystal, HOSEMANN [5] refers to the quantity $\sigma_c / \bar{d}_c$ as "g-factor".

If we knew that the long periods are varying from stack to stack, but not within one and the same stack, the quantities

$$\bar{L} = \frac{\mu'_1(h_L)}{\mu'_0(h_L)}$$

(average long period) and

$$\frac{\sigma_L}{\bar{L}} = \frac{\sigma(h_L)}{\bar{L}}$$

(relative standard deviation of the long periods, which is another HOSEMANN g-factor) describe the polydispersity of this material.

# 2 General Background

Interpretation of scattering data requires understanding of the general dimensions of the field and a general background of scattering theory which is reviewed in this chapter. Reference is given to textbooks and original work, where detailed discussion would extend beyond the scope of this book.

## 2.1 The Subareas of X-Ray Scattering

Scattering experiments are carried out in four different angular regions which will be frequently addressed in this book. In Table 2.1

the subareas are identified by the typical distance $R$ between the sample and the detector. The wavelength selected for the example is close to the historical wavelength of an X-ray tube equipped with a copper anode (CuK$_\alpha$ radiation with $\lambda = 0.15418$ nm).

Classical X-ray diffraction and scattering is carried out in the subarea of wide-angle X-ray scattering (WAXS). The corresponding scattering patterns yield information on the arrangement of polymer-chain segments (e.g., orientation of the amorphous phase, crystalline structure, size of crystals, crystal distortions, WAXS crystallinity).

The subarea of middle-angle X-ray scattering (MAXS) covers the characteristic scattering of liquid-crystalline structure and rigid-rod polymers.

In the small-angle X-ray scattering (SAXS) regime the typical nanostructures (in semicrystalline materials, thermoplastic elastomers) are observed. Because of the long distance between sample and detector time-resolved measurements can only be carried out at synchrotron radiation sources (Sect. 4.2.1.2).

**Table 2.1.** Subareas of scattering as a function of the sample–detector distance $R$ assuming an X-ray wavelength of $\lambda \approx 0.15$ nm

| Subarea | $R$ [m] | Focus |
|---------|---------|-------|
| WAXS | 0.05 – 0.2 | arrangement of chain segments |
| MAXS | 0.2 – 1 | liquid-crystalline structure |
| SAXS | 1 – 3 | nanostructure 3 nm – 50 nm |
| USAXS | 6 – 15 | nanostructure 15 nm – 2 $\mu$m |

The ultra small-angle X-ray scattering (USAXS) extends the accessible structure towards the micrometer range. Time-resolved measurements require a synchrotron beam that is intensified by an insertion device (Sect. 4.2.2).

## 2.2 X-Rays and Matter

### 2.2.1 General

X-rays are electromagnetic radiation with short wavelengths of about 0.01 to 10 nm. $\lambda \approx 0.15$ nm is the typical wavelength for the study of soft condensed matter. Whenever X-rays are interacting with matter, their main partners are the electrons in the studied sample. Thus X-ray scattering is probing the distribution of electron density, $\rho(\mathbf{r})$, inside the material.

As scattering intensity is computed from $\rho(\mathbf{r})$ in this book, the symbol $\rho(\mathbf{r})$ has two different meanings. Only in the field of WAXS it is identical to the plain *electron density*. However, in the area of SAXS it indicates the *electron density difference*[1], i.e., the deviation of the local electron density from the average electron density $\langle \rho(\mathbf{r}) \rangle_V$ in the irradiated volume $V$.

**Electron Density Computation.**   The average[2] electron density of a material or of a specific phase within a material,

$$\rho = Z_m \rho_m = N_A \frac{Z_M}{M_M} \rho_m, \tag{2.1}$$

is computed from the respective average mass density, $\rho_m$, by multiplication with the "number of electrons per gram", $Z_m$, given by Avogadro's number, $N_A = 6.022 \times 10^{23} \mathrm{mol}^{-1}$, the number of electrons per molecule or monomer unit, $Z_M$, and the molecular weight of molecule or monomer unit, $M_M$.

For polybutadiene with the chemical composition $C_4H_6$ we have a molecular weight of $M_M = 54.092\,\mathrm{g/mol}$ and $Z_M = 30\,\mathrm{e.u.}$ (electrons in "electron units"). If the mass density is $\rho_m = 0.90\,\mathrm{g/cm^3}$, the electron density becomes $\rho = 300.6\,\mathrm{e.u./nm^3}$.

### 2.2.2 Polarization

Polarization is a relevant issue, because we are dealing with transversal waves (GUINIER [6], p. 10-11). Polarization correction should be carried out for MAXS and WAXS data. It is less important for SAXS and USAXS patterns. In particular, if synchrotron radiation is used, the polarization correction is quite involved and based on the degree of polarization. For the purpose of reliable correction it is thus recommended to let a polarization monitor measure the actual degree of synchrotron beam polarization.

---

[1] In many publications the electron density difference is addressed as $\Delta\rho(\mathbf{r}) = \rho(\mathbf{r}) - \langle \rho(\mathbf{r}) \rangle_V$.

[2] *Exercise:* Compute the average electron density $\langle \rho \rangle_V$ of a sample from pure poly(ethylene terephthalate) (PET) with a mass density of 1.38 g/cm³. The chemical formula of PET is $C_{10}H_8O_4$. Because PET is most probably in the semicrystalline state, it makes sense to stress that the computed electron density is a volume average $\langle \rangle_V$.

### 2.2.2.1 Polarization Factor of a Laboratory Source

The polarization factor of a common X-ray source emitting unpolarized monochromatic light is

$$f_{P0}(2\theta) = \frac{1}{2}\left(1 + \cos^2(2\theta)\right) \tag{2.2}$$

a well-known function of the scattering angle $2\theta$ (ALEXANDER [7], p. 40; GUINIER [6], p. 99). Special care must be taken, if the monochromator is installed not in the primary beam (primary monochromatization), but in the diffracted beam (secondary monochromator) [8,9]. See also Sect. 2.2.3.

### 2.2.2.2 Synchrotron Beam Polarization Factor

Synchrotron light is, in general, polarized in horizontal direction ([10], p. 9-13). Nevertheless, the polarization of the beam is never perfect. In order to be able to carry out a quantitative polarization correction, the quality of polarization should be monitored by means of a polarization monitor [11] that is positioned in the primary beam. The polarization monitor is registering the horizontally polarized component, $I_h$, and the vertically polarized component, $I_v$. From these two intensities the quality

$$Q_P = \frac{I_h - I_v}{I_h + I_v} \tag{2.3}$$

of horizontal polarization is computed. In the ideal case $Q_P = 1$ is valid. Polarization monitors are rarely available in the field of elastic scattering. Thus, if polarization correction is carried out, it is frequently assumed that polarization is ideal horizontal. For some synchrotron beamlines the quality of polarization is part of the technical specification and can be queried. Values found in the worldwide web range between $Q_P = 0.95\ldots 1.00$ with typical uncertainties of 0.02.

From simple geometrical consideration (cf. Fig. 2.1) it follows that at a synchrotron the ideal polarization factor is only a function of the *horizontal* component of the scattering angle. Thus we conveniently express the polarization factor in terms of the horizontal scattering-angle

$$\theta_1 = \arctan(x_1/R)$$

and of the vertical scattering-angle

$$\theta_3 = \arctan(x_3/R)$$

in the nomenclature from Fig. 2.1. For the practically relevant polarization factor of synchrotron radiation

$$f_P(\theta_1, \theta_3) = \frac{(1 - Q_P)\cos^2\theta_3 + (1 + Q_P)\cos^2\theta_1}{2} \tag{2.4}$$

is obtained. Here $(1 - Q_P)/2$ and $(1 + Q_P)/2$, respectively, are the vertically and horizontally polarized fractions of the radiation. The reader easily verifies that the

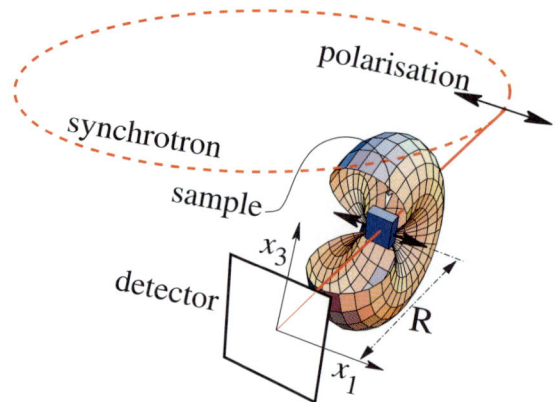

**Figure 2.1.** Polarization of synchrotron radiation in a hypothetic non-relativistic case. In reality the toroidal radiation pattern is degenerated to a narrow forward lobe

equation returns the correct solution for the ideal case ($Q_P = 1$). If both the rows of the *two-dimensional* (2D) detector and the primary beam are not tilted with respect to the plane of the storage ring and, moreover, the polarization is ideal the polarization correction becomes simple: all the intensities from the same column of detector pixels are divided by the same factor $f_P(\theta_1)$. A more involved treatment of polarization correction has been published by KAHN et al. [12].

### 2.2.3 Compton Scattering

An elimination of Compton scattering (also called incoherent scattering) should be carried out for WAXS data before quantitative evaluation. It is unnecessary for SAXS and USAXS patterns. Compton scattering is a result of an energy transfer from the photon to the electron. The intensity of the Compton scattering as a function of the scattering angle is computed from the chemical composition of the sample ([7] p. 29-32) ([13] p. 247-253). For application in the field of soft condensed matter it has been demonstrated by RULAND [14] that both the BREIT-DIRAC recoil-term and proper absorption correction of the Compton scattering should be considered for WAXS data. This specific absorption correction should not be confused with the general correction of the scattering pattern for absorption, which is discussed in Sect. 7.6 on page 76. If a secondary monochromator is installed[3], part of the Compton scattering will be eliminated by it [8].

### 2.2.4 Fluorescence

Fluorescence effects are, in general, considered to be negligible. In gas-filled detectors the corresponding photons are discarded in the energy discrimination stage of

---

[3]meaning that the device is installed in the diffracted beam

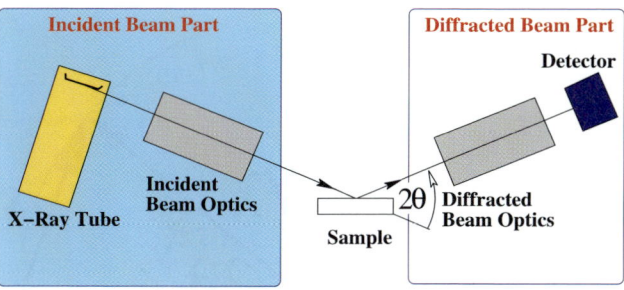

**Figure 2.2.** The classical setup for the static measurement of X-ray scattering in symmetrical-reflection geometry. $2\theta$ is the scattering angle. If the sample is turned upright and thus transmitted instead, the corresponding geometry becomes the symmetrical-transmission geometry

the electronics. Modern CCD-detectors, on the other hand, are sensitive to fluorescence photons.

## 2.3 Classical X-Ray Setup

In principle every scattering pattern can be recorded using the classical X-ray diffraction setup sketched in Fig. 2.2. In the detector the scattering intensity is measured in units of counts-per-second.

Using the ideal instrument we would vary the scattering angle $2\theta$ to record a scattering curve, then rotate and tilt[4] the sample in order to obtain the *complete* scattering pattern. In fact, these three angles are advantageously mapped to a *reciprocal space* that is inversely related to the *real space* in which the sought-after structure of the sample is defined.

## 2.4 s-Space and q-Space

In Fig. 2.3 the relation between the setting of the instrument and the actual position that is sensed in reciprocal space is sketched in the plane of incidence, i.e., in the plane that is spanned by two vectors $\mathbf{S_0}$ and $\mathbf{S}$ which are unit vectors[5] which indicate the directions of the incident and the scattered beam, respectively. Upon variation of the scattering angle $2\theta$, the tip of the scattering vector $\mathbf{s}$ is moving along a circular arc. Its magnitude

$$|\mathbf{s}| = s = \frac{2}{\lambda} \sin\theta \tag{2.5}$$

equaling the base of the isosceles triangle spanned by the congruent sides $\mathbf{S_0}$ and $\mathbf{S}$. If, in addition, the plane of incidence is rotated about the direction of the incident

---

[4]For a sketch cf. p. 193, Fig. 9.3

[5]The magnitude of a unit vector is 1.

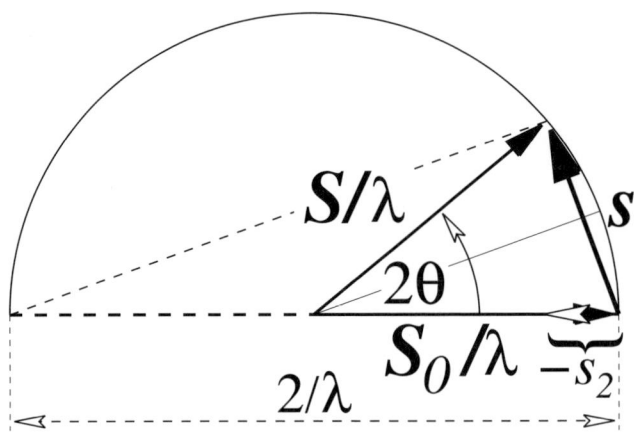

**Figure 2.3.** The scattering vector **s** probing the surface of the Ewald sphere upon variation of the scattering angle $2\theta$. $\lambda$ is the wavelength of the X-rays, and $s_2$ is the out-of-plane component of the scattering vector **s**

beam, **s** is probing the surface of the so-called Ewald sphere. Thus the scattering intensity $I(\mathbf{s})$ is considered a function of the scattering vector. **s** has three components which define the reciprocal space. Modern setups are frequently equipped with a plane 2D detector[6]. It should be clear that – because of the geometrical construction scheme – the image on the detector is not the representation of a plane in **s**-space. Instead, it represents the scattering intensity on a spherical[7] calotte of reciprocal space. The curvature of the detection surface is no problem at small angles (SAXS, USAXS), because here it is negligible – and the so-called tangent-plane approximation is valid with

$$\sin\theta \approx \tan\theta \approx \theta. \tag{2.6}$$

In practice, for SAXS and USAXS experiments carried out in normal-transmission geometry the set of equations

$$s_x = \frac{x}{\lambda R}, \quad s_y = \frac{y}{\lambda R}. \tag{2.7}$$

relates the position of each pixel $(x, y)$ on the detector as measured from the center of the scattering pattern to two components $s_x$ and $s_y$ of the scattering vector (cf. Sect. 10.1.2).

Because the orientation of the reciprocal space coordinate system is rigidly coupled to the orientation of the real-space coordinate system of the sample, the reciprocal space can be explored[8] by tilting and rotating the sample in the X-ray beam (cf. Chap. 9).

---

[6]Such detectors are frequently charge-coupled devices (CCD) known from digital cameras
[7]Ewald sphere!
[8]up to the distance defined by the diameter of the Ewald sphere

Two different definitions of the scattering vector, with

$$\mathbf{q} = 2\pi \mathbf{s} \tag{2.8}$$

are presently used in the field of scattering. In the past other letters ($\mathbf{m}$, $\mathbf{b}$) have been used as well, but there were never more than two different mathematical definitions. In the $\mathbf{s}$-system the relation between physical and reciprocal space is perfectly anti-symmetrical. In the $\mathbf{q}$-system this intrinsic mathematical symmetry has unnecessarily been broken [15]. For instance, a preferential distance $L$ between two particles results in a maximum at $s_L = 1/L$ but at $q_L = 2\pi/L$ according to Bragg's law.

According to the Fraunhofer approximation of kinematic scattering theory the real space and the reciprocal space are related to each other by an integral transform known by the name *Fourier transform*, which shall be indicated by the operator $\mathscr{F}\,()$. The *n-dimensional* (nD) Fourier transform of $h(r)$ is defined by

$$\mathscr{F}_n(h)(\mathbf{s}) := \int h(\mathbf{r}) \exp(2\pi i\mathbf{rs}) \, d^n r, \tag{2.9}$$

with $i$ the imaginary unit – and back-transformation simply yields

$$\mathscr{F}_{-n}(H)(\mathbf{r}) := \int H(\mathbf{s}) \exp(-2\pi i\mathbf{rs}) \, d^n s, \tag{2.10}$$

with $H(\mathbf{s}) := \mathscr{F}_n(h)(\mathbf{s})$. In the field of scattering 1D-, 2D- and 3D-transforms are required. The kernel of the Fourier transform is called the harmonic function

$$\exp(2\pi irs) = \cos(2\pi rs) + i\sin(2\pi rs), \tag{2.11}$$

and the Fourier transform is said to perform an harmonic analysis. $\mathfrak{R}()$ and $\mathfrak{I}()$ define the real and the imaginary part of a complex number, respectively.

In the $\mathbf{q}$-system one has asymmetrical transformation pairs made from

$$\mathscr{F}_n(h)(\mathbf{q}) := \int h(\mathbf{r}) \exp(i\mathbf{rq}) \, d^n r \tag{2.12}$$

and

$$\mathscr{F}_{-n}(H)(\mathbf{r}) := \left(\frac{1}{2\pi}\right)^n \int H(\mathbf{q}) \exp(-i\mathbf{rq}) \, d^n q. \tag{2.13}$$

Thus calculus is kept simple in the $\mathbf{s}$-system.

## 2.5 Scattering Intensity and Sample Structure

This section is devoted to the explanation of Eq. (2.14).

The fundamental relations between the electron density distribution inside the sample, $\rho(\mathbf{r})$, and the observed scattering intensity, $I(\mathbf{s})$ are conveniently combined in a sketch

$$\rho\left(\mathbf{r}\right) \overset{\mathscr{F}_3}{\Leftrightarrow} A\left(\mathbf{s}\right)$$
$$\star 2 \Downarrow \qquad \Downarrow ||^2 \tag{2.14}$$
$$P\left(\mathbf{r}\right) \underset{\mathscr{F}_3}{\Leftrightarrow} I\left(\mathbf{s}\right)$$

which is known by the name *magic square of scattering* (Eq. (2.14)). In particular in the field of SAXS the *correlation function*,

$$\gamma(\mathbf{r}) = \rho^{\star 2}(\mathbf{r})/\rho^{\star 2}(0) = P(\mathbf{r})/\rho^{\star 2}(0), \tag{2.15}$$

is a synonym for the *Patterson function* $P(\mathbf{r})$.

### 2.5.1 Lay-Out of the Magic Square

In the corners of the square we find functions. These functions describe the structure of our material.

Along the edges of the square there are mathematical operations. The Fourier transform describes the relation between the left and the right side of the square. Thus, on the left side we find the functions of physical space, and the reciprocal space is found on the right side. Double-headed arrows show that the path from the left to the right side is reversible. Unfortunately, reversion is impossible after we have moved from the top to the bottom of the square – and the scattering intensity $I(\mathbf{s})$ is located in the lower right corner of the square.

### 2.5.2 Analysis Options – Example for SAXS Data

Options of data analysis can be deduced from the magic square and our notions concerning the structure. As an example let us consider the case of small-angle X-ray scattering. Here it is, in general, assumed that the structure is described by a continuous density function. Although there is no[9] way back from intensity to density, there are several options for data analysis:

1. Utilize theory and find out, how some structure parameters can be determined from the intensity directly,

2. walk from the intensity along the lower edge half-way back to real space, where the transformed data are closer to human perception,

3. model a structure and fit it to the intensity or

4. in addition to item 2 carry out "edge enhancement" in order to visualize structure by means of the chord distribution function (CDF), $z(\mathbf{r})$, and interpret or fit it.

---

[9]Except for the case of anomalous SAXS, Sect. 8.9

The operation related to item 4 is displayed in an extended square

$$\rho\,(\mathbf{r}) \overset{\mathscr{F}_3}{\Leftrightarrow} A\,(\mathbf{s})$$

$$\star 2 \Downarrow \qquad \Downarrow |\;|^2\,. \qquad\qquad (2.16)$$

$$z\,(\mathbf{r}) \overset{\Leftrightarrow}{\underset{\Delta}{}} P\,(\mathbf{r}) \overset{\Leftrightarrow}{\underset{\mathscr{F}_3}{}} I\,(\mathbf{s})$$

One major goal of this book is to demonstrate the application of these options for data analysis.

**Effort of Data Analysis.** The mentioned options are listed in the order of increasing complexity for the scientist. When scattering curves (isotropic data) shall be analyzed, all the four listed options have proven to be manageable by many scientific groups.

In contrast, a real challenge is the analysis of scattering images from anisotropic materials, and in this subarea many scientists surrender and resort to the interpretation of peak positions and peak widths in raw data (cf. citation of P. Debye on p. 1). So after having advanced by learning how to analyze curves, in the field of anisotropic materials we are now in a similar situation as science has been in 1931 in respect to isotropic data.

A shortcut solution for the analysis of anisotropic data is found by mapping scattering images to scattering curves as has been devised by BONART in 1966 [16]. Founded on Fourier transformation theory he has clarified that information on the structure "in a chosen direction" is not related to an intensity curve sliced from the pattern, but to a projection (cf. p. 23) of the pattern on the direction of interest.

The barrier to the application of the shortcut is probably resulting from the need to preprocess the scattering data and to project the 3D scattering intensity on a line. This task requires 3D geometrical imagination[10] and knowledge of methods of digital image processing, a field that is quite new to the community of scatterers. Programmers, on the other hand, are rarely educated in the fields of scattering and multidimensional projections.

### 2.5.3 Parameters, Functions and Operations in the Magic Square

$V$        is the irradiated volume. It is defined by the sample thickness multiplied by the footprint of the incident primary beam on the sample.

$\rho\,(\mathbf{r})$        the electron density (WAXS) or the electron density difference (SAXS) (cf. Sect. 2.2.1).

$A\,(\mathbf{s})$        scattering amplitude

$k = \rho^{\star 2}\,(0)$    scattering power

---

[10]Lack of 3D geometrical imagination and unawareness of the related mathematics is a frequent reason for widespread malpractice.

$Q = k/V$     invariant

$P(\mathbf{r})$          Patterson function

$\gamma(\mathbf{r}) = P(\mathbf{r})/k$ SAXS correlation function

$z(\mathbf{r})$          chord distribution function (CDF)

The X-ray detector measures the intensity of electromagnetic waves, i.e., the absolute square $|\ |^2$ of their amplitude. Thus, in combination, the upper path between density and intensity through the square is written as

$$I(\mathbf{s}) = |\mathscr{F}_3(\rho(\mathbf{r}))|^2.$$

In the lower path through the square we have an equivalent formulation

$$I(\mathbf{s}) = \mathscr{F}_3(\rho^{*2}(\mathbf{r}))$$

with the Patterson or correlation function $\rho^{*2}(\mathbf{r})$ involved (DEBYE (1949) [17], POROD (1951) [18]). $\rho^{*2}(\mathbf{r})$ is generated from the "inhomogeneities" $\rho(\mathbf{r})$ by means of the autocorrelation operator that will immediately be introduced.

### 2.5.4 Convolution, Correlation and Autocorrelation

**Convolution.**   The convolution of two 1D functions $f(r)$ and $g(r)$ is defined by

$$h(r) = \int_{-\infty}^{\infty} f(y) g(r-y)\, dy \tag{2.17}$$
$$:= f(r) \star g(r). \tag{2.18}$$

The definition of convolution is readily extended to the $n$-dimensional case. A convolution is frequently used in many fields of science and in digital image processing (blurring, unsharp masking).

**Correlation.**   Similar to convolution the correlation operator is

$$h(x) = \int_{-\infty}^{\infty} f(y) g(x+y)\, dy \tag{2.19}$$
$$:= f(x) \otimes g(x) = f(x) \star g(-x). \tag{2.20}$$

If $f(x)$ and $g(x)$ are different, the integral is named cross-correlation. If both functions are identical, the integral is named autocorrelation. In the latter case we write

$$h^{*2}(r) = h(r) \star h(-r) = \int_{-\infty}^{\infty} h(y) h(r+y)\, dy. \tag{2.21}$$

In the field of scattering the autocorrelation is also known by the name "convolution square".

**Shape functions.**   A shape function

$$Y(\mathbf{r}) = \begin{cases} 1 \ /\mathbf{r} \text{ inside the region} \\ 0 \ /\mathbf{r} \text{ outside the region} \end{cases} \tag{2.22}$$

describes a region. A region is, for example, a particle, a microfibril, a spherulite, the silhouette of a person on a picture. In digital image processing shape functions are named masks or regions of interest (ROI). Shape functions are the basic elements of topological structure both in the fields of scattering and diffraction. The Fourier transform of a shape function is denoted by

$$\Phi(\mathbf{s}) = \mathcal{F}(Y(\mathbf{r})). \tag{2.23}$$

**Convolution: Illustration.**   By convolution with the $\delta$-function $\delta(\mathbf{r} - \mathbf{r}')$ (cf. p. 25) we displace (translate) the particle by $\mathbf{r}'$. If, for the purpose of particle translation by convolution, we employ an abstract one-dimensional lattice, i.e.,

$$c_a(r) = \sum_{k=-\infty}^{\infty} \delta(r - ka), \tag{2.24}$$

the result will be a real lattice generated by cloning the particle infinitely and placing the clones at equal distances of $a$. This is the fundament of diffraction and crystallography. The principle is readily extended to more than one dimension. In the field of digital image processing and in the general theory of measurement[11] the function $c_a(r)$ is known by the name comb function. The Fourier transform of a comb function $C(s) = \mathcal{F}(c_a(r))(s)$ is, again a comb function

$$C(s) = \frac{1}{a} c_{1/a}(s). \tag{2.25}$$

**Autocorrelation: Illustration.**   We choose a shape function $Y(\mathbf{r})$ which describes a particle in 2D space (cf. Fig. 2.4a). Because of the definition of $Y(\mathbf{r})$, $Y^{*2}(\mathbf{r})$ takes the value of the volume which is shared by the particle and its imagined "ghost" which is displaced by $\mathbf{r}$. In any case the overlap integral becomes maximal for $\mathbf{r} = 0$. Here the correlation is perfect.

The autocorrelation operation does neither affect

$$\delta^{*2}(r) = \delta(r) \tag{2.26}$$

the $\delta$-function nor the principal shape[12] of the comb function

$$\lim_{N \to \infty} \left( \frac{1}{N} c_{Na}^{*2}(r)/N \right) = c_a(r), \tag{2.27}$$

---

[11] The abstract principle of measurement is multiplication of a continuous function by a comb function returning a series of discrete values.

[12] $N$ is the number of $\delta$-peaks in the comb function.

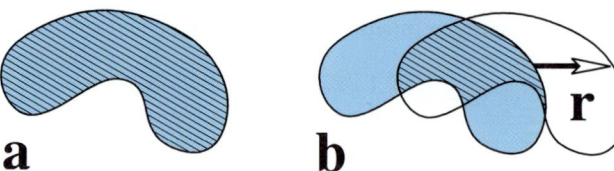

**Figure 2.4.** Illustration of the autocorrelation of some particle described by a function $Y(\mathbf{r})$. (a) $Y^{\star 2}(0)$: The particle and its imagined ghost are on top of each other. The overlap integral (hatched) is maximal. (b) $Y^{\star 2}(\mathbf{r})$: The ghost has been displaced by $\mathbf{r}$ with respect to the particle. The hatched area has decreased

with

$$c_{Na}(r) = \sum_{k=(1-N)/2}^{(N-1)/2} \delta(r - ka)$$

and

$$c_{Na}^{\star 2}(r) = \sum_{k=(1-N)}^{(N-1)} (N - |k|)\, \delta(r - ka).$$

These are two important special cases. The power and simplicity of diffraction pattern analysis (crystallography) for the analysis of regular structure is a result of Eq. (2.27) and Eq. (2.25). No information is lost if infinite abstract lattices are subjected to Fourier transformation.

The general case of scattering is less favorable. The decrease of the correlation function with increasing $\mathbf{r}$ depends both on the shape of the particle itself and on the arrangement of neighboring particles. In principle the maximum information of a scattering pattern is such correlation information.

## 2.6 Polydispersity and Scattering Intensity

Polydispersity is one of the most frequent reasons that soft condensed matter does not show diffraction but scattering. Thus its consideration is of utmost importance. The general effect of polydispersity on scattering patterns is demonstrated in this section.

Let us consider a *template*, i.e., the average representative particle or the average representative structural entity in a material with polydisperse structure. The template is described by its structure $\rho_T(\mathbf{r})$. The sample is full of dilated images

$$\rho_i(\mathbf{r}) = \rho_T\left(\frac{\mathbf{r}}{\mathbf{a}}\right) \tag{2.28}$$

of the template, but there is no arrangement (correlation) among these images. Here $\mathbf{a}$ is a scaling vector. In Cartesian coordinates it is written $\mathbf{a} = (a_1, a_2, a_3)$, and all its components (the scaling factors) $a_i \in (0, \infty)$ are positive. In Eq. (2.28) the scaling division is defined according to BRYCHKOV [19] $\mathbf{r}/\mathbf{a} := (r_1/a_1, r_2/a_2, r_3/a_3)$ in an unusual but in this context natural way[13].

The scattering intensity of the template

$$I_T(\mathbf{s}) = \int_{-\infty}^{\infty} \rho_T^{\star 2}(\mathbf{r}) \exp(2\pi i \mathbf{r}\mathbf{s}) \, d^3 r$$

$$:= \mathscr{F}\left(\rho_T^{\star 2}(\mathbf{r})\right)(\mathbf{s})$$

is the Fourier transform of $\rho_T^{\star 2}(\mathbf{r})$. For any of the images its scattering intensity is readily established utilizing two basic theorems of Fourier transformation theory[14] concerning dilation (in one dimension: $\mathscr{F}(aH(as)) = h(r/a)$, cf. p. 24) and convolution (cf. p. 25). Thus, in Cartesian coordinates and with the definition of scaling multiplication $\mathbf{a} \bullet \mathbf{s} := (a_1 s_1, a_2 s_s, a_3 s_3)$,

$$I_i(\mathbf{s}) = a_1^2 a_2^2 a_3^2 I_T(\mathbf{a} \bullet \mathbf{s}) \qquad (2.29)$$

is the scattering intensity of the image (dilated template). In particular for the isotropic case with isotropic dilation $a_i = a \; \forall \; i \in [1,2,3]$ we obtain the well-known result [21]

$$I_i(\mathbf{s}) = a^6 I_T(a\mathbf{s}), \qquad (2.30)$$

which is frequently cited when the meaning of the Guinier radius (Sect. 8.1) of a polydisperse material is discussed by referring to the distorting multiplication by a high power of the scaling factor. The less distorting effects of uniaxial dilation ($a_1 = a_2 = 1, a_3 = a$) and lateral dilation ($a_1 = a_2 = a, a_3 = 1$) on the scattering intensity of the dilated particle are readily established. The cases of isotropic, lateral and uniaxial dilation are the most important ones in the field of polydisperse structure.

There shall be no correlation among different structural entities. Thus the observed correlation function of the material

$$\rho^{\star 2}(\mathbf{r}) = \iiint h_H(\mathbf{a}) \, \rho_T^{\star 2}\left(\frac{\mathbf{r}}{\mathbf{a}}\right) d^3 a \qquad (2.31)$$

is a superposition of dilated correlation functions with $h_H(\mathbf{a})$ the size distribution of the structural entities. Determination of this size distribution is the aim of research in the field of polydisperse materials.

From the observed correlation function the scattering pattern is obtained by Fourier transformation. As Eq. (2.31) is subjected to the Fourier transform, it will only act on the correlation function of the *template* because $h_H(\mathbf{a})$ is no function of $\mathbf{r}$. With Eq. (2.29) we obtain the expected result

$$I(\mathbf{s}) = \iiint_0^{\infty} a_1^2 a_2^2 a_3^2 h_H(\mathbf{a}) I_T(\mathbf{a} \bullet \mathbf{s}) \, d^3 a \qquad (2.32)$$

---

[13] A more elegant way to introduce polydispersity is founded on Tensor calculus. For an application in scattering theory cf. e.g. BURGER and RULAND [20].

[14] We follow a common notation and denominate functions in real space with lower case letters and their Fourier transforms by the corresponding upper case letters: $\mathscr{F}(g(\mathbf{r})) = G(\mathbf{s})$

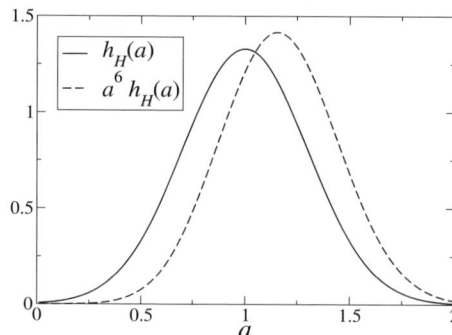

**Figure 2.5.** Polydispersity in the structure (*solid line*; Gaussian with $\sigma = 0.3$) and the resulting effective size distribution in the scattering intensity (*broken line*)

in Cartesian coordinates assuming the conservation of contrast[15]. Contrast is the electron density difference between the highest electron density in the template and the electron density of the surrounding matrix. Reduced to isotropic material and isotropic dilation we have

$$I(s) = \iiint_0^\infty a^6 h_H(a) I_T(as) d^3 a. \tag{2.33}$$

Resorting to Sect. 1.3 we thus find that in the scattering intensity there is no "natural" weighting of different monodisperse fractions of images inside a polydisperse material. The most frequent isotropic case is most severely affected. For the case of isotropic dilation Fig. 2.5 shows the distortion of a Gaussian[16] scaling factor distribution, $h_H(a)$, with a standard deviation $\sigma = 0.3$, which is a typical value for materials from synthetic polymers. In the scattering intensity the distorted distribution (broken line) is effective instead of the undistorted broad Gaussian (solid line).

**Averages of Structural Parameters.** The problem becomes less severe if we are no longer interested in the determination of the scaling factor distribution $h_H$ itself, but turn to the determination of averages of structural parameters by integration, instead. Let us consider the scattering of an ensemble of spheres. In the limit of a monodisperse system, we could directly determine the sphere diameter from the scattering curve. In the case of a polydisperse system, basic mathematics shows that the number average diameter, $2 \langle R \rangle_n$, is not accessible. Instead, the number average sphere volume

$$4\pi \int_0^\infty s^2 I(s) ds = \frac{4\pi}{3} \langle R^3 \rangle_n$$

---

[15] All images of the template are "compact" in the same manner. They do not become less dense upon expansion, as it would be the case if the mass were conserved.

[16] Normalized $d$-dimensional Gaussian $h(r) = \left(1/\left(2\pi\sigma^2\right)\right)^{d/2} \exp\left(-\left((r-r_0)^2/2\sigma^2\right)\right)$ at $r_0$ with standard deviation $\sigma$.

can be computed. Here we have assumed an electron density difference $\rho = 1$. Moreover, the intensity $I(s)$ is defined by the scattering per sphere in an infinitely diluted system. Thus not considering polydispersity would lead to a considerably overestimated sphere diameter.

The pedestrian approach to polydispersity that has been demonstrated up to here is an extension to three dimensions of the well-known rigorous treatment of polydispersity in one dimension by means of the Mellin convolution (Eq. (8.85), p. 168).

**The Tensor Approach to Polydispersity** is treating the problem on a much more universal level. Here it shall at least be sketched. Instead of Eq. (2.28) we write

$$\rho_i(\mathbf{r}) = \rho_T(\mathbf{T}_i\mathbf{r}) \tag{2.34}$$

with $\mathbf{T}_i$ a tensor which maps the average reference structure, $\rho_T(\mathbf{r})$, to the structure $\rho_i(\mathbf{r})$ of the i-th particle (of the polydisperse ensemble). According to the general theory of $n$-dimensional Fourier transformation, the corresponding scattering amplitude is

$$A_i(\mathbf{s}) = \frac{1}{|\mathbf{T}_i|} A_T\left(\left(\mathbf{T}_i^{-1}\right)^T \mathbf{s}\right),$$

with

$$A_T(\mathbf{s}) = \mathscr{F}(\rho_T(\mathbf{r}))$$

the Fourier transform of the template. $\left(\mathbf{T}_i^{-1}\right)^T$ denominates the transposition of the inverse of $\mathbf{T}_i$, and $|\mathbf{T}_i|$ is its determinant. The scattering intensity

$$I_i(\mathbf{s}) = \frac{1}{|\mathbf{T}_i|^2} |A_T|^2\left(\left(\mathbf{T}_i^{-1}\right)^T \mathbf{s}\right)$$

is the square of the amplitude. For the special case of dilation as treated above we have

$$\mathbf{T}_i = \begin{pmatrix} 1/a_{i,1} & 0 & 0 \\ 0 & 1/a_{i,2} & 0 \\ 0 & 0 & 1/a_{i,3} \end{pmatrix},$$

and many of the equations given above are readily established. In general, the tensor calculus is valid for all mappings which are described by a tensor with $|\mathbf{T}_i| > 0$. In particular orientation distributions (cf. Chap. 9) are thus as well covered by this tensor formalism.

## 2.7 A Glance at the Mathematical Laboratory of Scattering

Favorable properties of the Fourier transform itself provide general means either to split the general problem of data analysis into sub-problems or even to obtain structure parameters without much modeling work. In this respect the Fourier slice theorem must be pointed out because of its superior impact on scattering (BONART [16];

BALTÁ and VONK [22], p. 15) and on several modern technologies[17]. The theorem deals with projections and slices. It explains the weird information on structure that we retrieve if we study the scattering intensity cut from a pattern along a line that is extending outward from the center of the pattern. In fact, the respective intensity curve is called a slice (or a section). Last but not least, the theorem reveals an elegant way to overcome the recognized problem.

In combination with other theorems of Fourier transformation theory many of the fundamental structural parameters in the field of scattering are readily established. Because the corresponding relations are not easily accessible in textbooks, a synopsis of the most important tools is presented in the sequel.

### 2.7.1 The Slice

Generalizing the reasoning in the introduction above, we consider a deliberate function $f$. It has to be defined in space, but there is good reason not to specify if it is the real or the reciprocal one. So let us denote the space by a vector $\mathbf{u}_n$. The index $n$ shall indicate that this vector has $n$ components, i.e., $\mathbf{u}_n \in \mathbb{R}^n$ is a member of the $n$-dimensional vector space of real numbers over which the function $f$ shall take its values. For application we will identify $\mathbf{u}_n$ either by $\mathbf{s}$ (and then the function $f$ may be the scattering intensity $I(\mathbf{s})$), or by $\mathbf{r}$. Now let us consider a vector $\mathbf{u}_m \in \mathbb{R}^m$ that has less components ($m < n$) than $\mathbf{u}_n$. We say $\mathbb{R}^m$ is a subspace of $\mathbb{R}^n$. This may be a plane or a line through $\mathbf{s}$-space or $\mathbf{r}$-space, respectively. There are many possible subspaces, but for the slice theorem to work we have to choose special subspaces which include the origin of the coordinate system. If we restrict the considered $n$-dimensional function $f(\mathbf{u}_n)$ to an $m$-dimensional subspace $\mathbf{u}_m$, we indicate this by writing $\lceil f \rceil_m (\mathbf{u}_m)$ and call this restricted function the $m$-dimensional slice (or section) of $f$ in the coordinates $\mathbf{u}_m$. Obviously, we can always rotate the coordinate system of $\mathbf{u}_n$ in such a way, that the redundant coordinates $\mathbf{u}_{n-m} = 0$ become zero. So the general mapping rule of a slice is

$$\lceil f \rceil_m (\mathbf{u}_m) = f(\mathbf{u}_n)|_{\mathbf{u}_{n-m}=0} \tag{2.35}$$

with the vertical bar meaning "restricted to" or "at the position". So let us consider examples now. For a fiber scattering pattern $I(\mathbf{s})$ with the fiber axis rotated into $s_3$-direction,

$$\lceil I(\mathbf{s}) \rceil_1 (s_3) = I(0,0,s_3) \tag{2.36}$$

is an example for a 1D slice. It is a curve taken from the pattern along the meridian[18]. A different slice is

$$\lceil I(\mathbf{s}) \rceil_2 (s_1,s_2) = I(s_1,s_2,0) = \lceil I(\mathbf{s}) \rceil_2 (s_{12}).$$

---

[17]Computer tomography, magnetic resonance imaging, digital image processing, synchrotron micro-tomography [23,24], 3D electron microscopy of block copolymers [25].

[18]Meridian is the name for the principal axis found in the scattering patterns of uniaxial materials.

It is a 2D slice of the fiber pattern. Only fiber symmetry makes that it is completely represented by a curve as a function of a transversal (cf. BONART [16]) coordinate $s_{12} = \sqrt{s_1^2 + s_2^2}$ on the equator[19] of the pattern.

Resorting to the definition of the Fourier transform, Eq. (2.9), we notice that for the redundant coordinates the harmonic kernel degenerates and becomes $\exp(0) = 1$. Thus for the redundant coordinates the Fourier transform turns into a simple integration with respect to the respective reciprocal coordinate[20] – a "projection".

### 2.7.2 The Projection

As with the slice, the projection, as well, is a mapping of a function $f(\mathbf{u}_n)$, $\mathbf{u}_n \in \mathbb{R}^n$, to a subspace. Its mapping rule is

$$\{f\}_m (\mathbf{u}_m) = \int f(\mathbf{u}_n)\, d\mathbf{u}_{n-m}. \tag{2.37}$$

Thus we integrate $f$ over all those Cartesian coordinates from which the projected curve shall no longer be a function. Unexperienced scientists tend to make mistakes by simply "summing pixels" from the 2D scattering image collected on the detector, although the problem is from 3D $\mathbf{s}$-space. So the question to answer first is: is the information that I have gathered in my experiment complete? If the affirmative answer has been justified, the means of how the integration has to be performed are right at hand.

### 2.7.3 Fourier Slice Theorem

Under Fourier transform, slice and projection are exchanged and it follows

$$\lceil h \rceil_m = \mathscr{F}^m (\{H\}_m), \tag{2.38}$$

with $h(\mathbf{r})$ some function and $H(\mathbf{s})$ being its Fourier transform. The slice theorem is also known by the name *central projection theorem*.

### 2.7.4 Fourier Derivative Theorem

From the definition of Fourier transform the derivative theorem

$$\mathscr{F}\left(\frac{d^n h(r)}{dr^n}\right) = (2\pi i s)^n H(s) \tag{2.39}$$

is established by partial derivation. Extension to the multidimensional case is simple for even orders of the derivative (STRIBECK (2001) [26])

$$\mathscr{F}\left(\nabla^{2n} h(\mathbf{r})\right) = \left(-4\pi^2 s^2\right)^n H(\mathbf{s}) \tag{2.40}$$

$$\mathscr{F}\left(\Delta h(\mathbf{r})\right) = -4\pi^2 s^2 H(\mathbf{s}), \tag{2.41}$$

---

[19]Equator is the name of the direction perpendicular to the principal axis found in the scattering patterns of uniaxial materials.

[20]The respective reciprocal coordinate is called "dual coordinate" by the mathematician.

with $\nabla$ the gradient ("nabla") operator and

$$\Delta = \sum_i \frac{\partial^2}{\partial u^2} \tag{2.42}$$

the Laplacian – with the given definition valid in Cartesian coordinates.

### 2.7.5 Breadth Theorem

The integral breadth of a 1D, even and Fourier-transformable function $h(r)$ is defined by

$$B(h) = \frac{\int h(r)\, dr}{h(0)}. \tag{2.43}$$

Then it follows from the slice theorem Eq. (2.38) for the integral breadth of the Fourier transformed function $H(s)$

$$B(H) = \frac{\int H(s)\, ds}{H(0)} = \frac{h(0)}{\int h(x)\, dx} = \frac{1}{B(h)}. \tag{2.44}$$

In the field of scattering a simplified version of the Fourier breadth corollary Eq. (2.44) is known as the SCHERRER equation[21]. As a result, the inverse of the *integral* breadth of a peak or reflection is the size of the crystal in the direction perpendicular to the netplanes that are related to the reflection.

In order to deduce SCHERRER's equation first an infinite crystal is considered that is, second, restricted (i.e multiplied) by a shape function (cf. p. 17). Thus from the Fourier convolution theorem (Sect. 2.7.8) it follows that in reciprocal space each reflection is convolved by the Fourier transform of the square of the shape function – and SCHERRER's equation is readily established.

### 2.7.6 Dilation and Reciprocity

From the definition of the Fourier transform it follows that

$$\mathscr{F}\left(\frac{1}{a} h\left(\frac{r}{a}\right)\right) = H(as). \tag{2.45}$$

It is worth to be noted that $(1/a)\, h(r/a)$ is the result of the dilation of $h(r)$ by the factor $a$ in which the area under the curve is conserved. The result in reciprocal space is a compressed function $H$. This property of the Fourier transform is the generalization of Bragg's law.

---

[21] In the literature the SCHERRER equation is frequently related to the full widths at half-maximum. This approximation is unnecessary.

### 2.7.7 DIRAC's $\delta$-Function

A definition of DIRAC's $\delta$-distribution is readily established from dilation. Let

$$\int_{-\infty}^{\infty} h(r)\, dr = 1 \tag{2.46}$$

be normalized. Then the dilated function

$$\int_{-\infty}^{\infty} h\left(\frac{r}{a}\right) \frac{dr}{a} = 1 \tag{2.47}$$

is still normalized and $\delta(r)$ is defined taking limits

$$\delta(r) = \lim_{a \to 0} \frac{1}{a} h\left(\frac{r}{a}\right) \tag{2.48}$$

whereupon the integral remains normalized.

### 2.7.8 Convolution Theorem

Under Fourier transform the convolution (Eq. (2.17)) is turned into a multiplication.

$$\mathscr{F}(f \star g) = F\,G \tag{2.49}$$
$$\mathscr{F}(f g) = F \star G. \tag{2.50}$$

This property is readily established from the definition of Fourier transform and convolution. In scattering theory this theorem is the basis of methods for the separation of (particle) size from distortions (STOKES [27], WARREN-AVERBACH [28,29]: lattice distortion, RULAND [30–34]: misorientation of anisotropic structural entities) of the scattering pattern.

### 2.7.9 Bandlimited Functions

If, in practice, a Fourier transformation shall be carried out, it is meaningful to search for functions that are not only bounded, but, which even vanish when taking limits $|\mathbf{s}| \to \infty$ or $|\mathbf{r}| \to \infty$. Such functions are called bandlimited. Let us consider the function $h(\mathbf{r}) = \mathscr{F}(H(\mathbf{s}))$. Then the reciprocal space image $H(\mathbf{s})$ is bandlimited if its Fourier transform, $h(\mathbf{r})$, does not contain spatial frequencies above a value of $f_u$, i.e.,

$$h(\mathbf{r}) = 0 \text{ for } |\mathbf{r}| > f_u , \tag{2.51}$$

and $f_u$ is the upper frequency of the frequency band. In mathematics, band limitation is expressed in terms of functions with "finite support". The support

$$\mathrm{supp}\,(h) \tag{2.52}$$

of the function $h$ is the region, in which the function $h(\mathbf{r})$ does *not* vanish.

## 2.8 How to Collect Complete Scattering Patterns

Resorting to Debye (cf. p. 1), *"only a continuous scattering pattern can be the fundament of proper reasoning"* the general question must be addressed, how a complete scattering pattern can be collected. The considerations of this section are based on the assumption that the scattering pattern is recorded by means of a 2D- or 1D-detector.

### 2.8.1 Isotropic Scattering

**The Limits.**    There are a lot of materials whose scattering pattern does not change if the sample is deliberately rotated in the X-ray beam. Such materials are called isotropic. For isotropic materials completeness is only a question of the angular range in which significant scattering information is gathered. The technical limits are defined by the setup, and the fundamental parameter is the distance $R$ between sample and detector. The smallest accessible scattering angle is given by the size of the beam stop (cf. p. 37, Fig. 4.1b) which prevents the detector from being damaged by the direct beam. The highest angle with reasonable data is restricted by the extension of the detector or, worse, by the signal-to-noise (S/N) ratio of the data. If thin samples are exposed for short time in a weak beam, there is most probably no significant information in the outer part of the scattering pattern and quantitative data evaluation is futile. The problem is less severe if a 2D-detector is used. In this case azimuthal averaging will increase the S/N-ratio in particular at high scattering angles.

**How to Arrange the Setup.**    In practice, the distance $R$ is long enough, if the scattering intensity can safely be extrapolated towards zero from the data recorded. The distance $R$ is short enough, if in the outer part of the scattering pattern, a sufficiently long region with a monotonous background is recorded. One should not underestimate the need for sufficient recording of background in SAXS and US-AXS. In order to increase the highest accessible angle, 2D detectors may be placed in a lateral off-set position with respect to the primary beam.

If there is no possibility to cover the complete range with one detector, there may be the possibility to use two detectors which are placed in different distances from the sample. In the worst case the experiment has to be performed several times with different setups.

### 2.8.2 Anisotropic Scattering

Anisotropy is frequently observed in soft materials, but the symmetry of anisotropy is varying. Fibers and films show, in general, less complex anisotropy than ordinary or photonic crystals.

### 2.8.2.1 Single Crystal Anisotropy

Complete scattering patterns of samples with a complex "single-crystal" anisotropy can only be recorded in a texture setup (Chap. 9, Fig. 9.3). The samples must be rotated in order to scan the required fraction of reciprocal space.

### 2.8.2.2 Fiber Symmetry

**Definition.**   Fiber symmetry is uniaxial or cylindrical symmetry. Revolving the sample about the fiber axis does not change the scattering pattern, but tilting the sample with respect to the fiber axis does.

**USAXS and SAXS.**   Concerning USAXS and SAXS, the scattering pattern that is recorded on a 2D detector is complete if the principal axis of the sample is set normal to the direction of the incident X-ray beam (primary beam). Completeness is a result of two facts.

1. Fiber symmetry: with the $s_3$ axis in fiber direction the pattern shows rotational symmetry in the plane $(s_1, s_2)$, thus $I(\mathbf{s}) = I\left(\sqrt{s_1^2 + s_2^2}, s_3\right) = I(s_{12}, s_3)$ is a function of $s_3$ and of the distance from this axis only.

2. The tangent plane approximation is valid: the curvature of the Ewald sphere is negligible at small scattering angles.

Thus in this favorable case the complete information on nanostructure is recorded in one 2D image. Mathematically the recorded image is a slice

$$\lceil I(\mathbf{s}) \rceil_2 (s_1, s_3) \equiv I(s_{12}, s_3). \tag{2.53}$$

It is complete because of fiber symmetry. The 2D Fourier transform of this image is not related to the searched slice, but to a *projection* of the correlation function. In contrast, the sought-after slice in real space

$$\rho^{\star 2}(r_{12}, r_3) = \lceil \rho^{\star 2}(\mathbf{r}) \rceil_2 (r_1, r_3)$$
$$= F^2(\{I(\mathbf{s})\}_2 (s_1, s_3)),$$

is the 2D Fourier transform of the projection

$$\{I(\mathbf{s})\}_2 (s_1, s_3) = \int I\left(\sqrt{s_1^2 + s_2^2}, s_3\right) ds_2$$

of the complete intensity from the 3D scattering pattern on the slice formed by the detector plane. Because of completeness it can be computed from the data collected in one 2D scattering pattern.

**WAXS and MAXS.**   Fiber symmetry means that, even in WAXS and MAXS, the scattering pattern is completely described by a slice in reciprocal space that contains the fiber axis. Nevertheless, for $2\theta > 9°$ the tangent plane approximation is no longer valid and the detector plane is mapped on a spherical surface in reciprocal space.

If we keep the sample's principal axis normal to the primary beam and record a scattering pattern, we can readily map the measured intensities to the plane that we need to know (BUERGER (1942) in ALEXANDER [7], p. 58-62). For this purpose

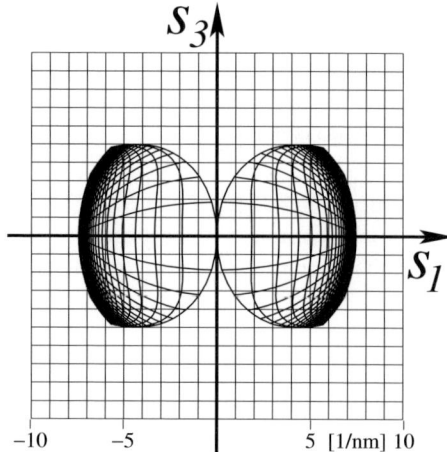

**Figure 2.6.** WAXS, 2D-detector and fiber symmetry: unwarping of the detector surface to map it on the $(s_1, s_3)$-slice. Fiber direction is normal to the primary beam. $R = 10$ cm, $\lambda = 0.154$ nm. The warped grid in the sketch is a square grid on the detector (edge length: 3 cm)

we refer to Fig. 2.3 and deduce the out-of-plane component $s_2$, which is readily established by application of Pythagoras' cathetus theorem[22]. Thereafter we compute the components $s_1$ and $s_3$ and receive the mapping equations. The result shows a peculiar deformation (Fig. 2.6). With respect to the slice that contains the complete information, only the area enclosed by solid lines is recorded on the plane detector. There are two blind gusset-shaped areas extending from the center upward and downward along the meridian. Within these areas Bragg peaks may be hidden. Thus the scattering pattern of fibers collected on the 2D detector is not complete if WAXS data are recorded.

It is worth to be noted that not only the position of the pixels, but also their area is modified by the unwarping. Correction of WAXS images thus requires both a translation and a magnification of the intensity proportional to the inverse of the area enclosed by the respective vertices. After the advent of digital computers it became possible to carry out the cumbersome calculus automatically[23], as proposed by FRASER[24] et al. [35].

The solution to access the invisible areas is readily copied from texture analysis: *tilt the sample* by $\psi$ and receive 1 data point on the meridian that corresponds to $s_3 = (2/\lambda) \sin \psi$. The result of the mapping is shown in Fig. 2.7. Thus by recording a series of images taken at different tilt angles of the fiber the blind area can be covered to a sufficient extent. Finally, the remnant blind spots may be covered by means of

---

[22] $(-2s_2/\lambda = s^2$ in the right triangle under THALES' circle whose leg is indicated by a dashed line). The use of the cathetus theorem was suggested by my daughter Agnes.

[23] A corresponding program was presented by RICHARD HILMER (DuPont Inc., Wilmington, USA) at a CCP13 workshop in 1997. The program is property of DuPont.

[24] B. HSIAO and scientists of his group have started to call the algorithm "Fraser correction"

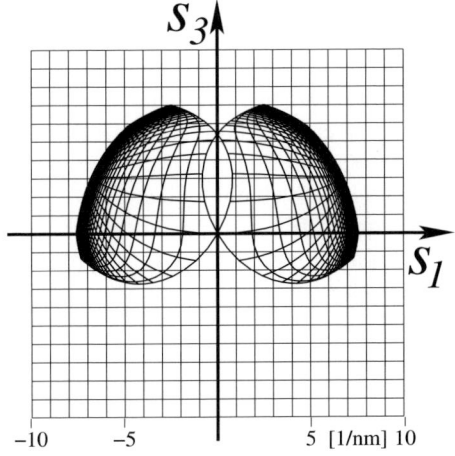

**Figure 2.7.** WAXS, 2D-detector and fiber symmetry: unwarping of the detector surface to map it on the $(s_1, s_3)$-slice. Fiber direction is tilted by $\psi = -30°$ with respect to the primary beam. $R = 10$ cm, $\lambda = 0.154$ nm. On the detector the apparent warped grid is a square grid (edge length: 3 cm)

2D extrapolation procedures, e.g., the algorithm based on radial basis functions [36] which is implemented in *pv-wave*® [37].

## 2.9 Application of Digital Image Processing (DI)

### 2.9.1 DI and the Analysis of Scattering Patterns

In 1994, when the bottleneck of scattering data analysis was still the poor performance of detectors, RUDOLPH & LANDES were already spotting the bottleneck of our days:

> "Having 2D detection that operates in the cycle time of key experiments means we are then potentially limited by image processing. In other words, as soon as we begin using 2D-detectors to measure patterns, we are forced to use image analysis methods to extract information from the images. With the rapid development of fast detectors, image analysis becomes key to our effective use of this technology."( [38], p. 26)

The source code of a set of DI procedures for the processing of scattering patterns written for *pv-wave* is available on the worldwide web (www.chemie.uni-hamburg.de/tmc/stribeck/dl/).

## 2.9.2 A Scattering Pattern Is a Matrix of Numbers, Not a Photo

Digital image processing starts after we have managed to read the raw scattering image. Each scattering pattern recorded by the detector is a matrix filled with positive integer numbers – the counts in each cell (pixel) of the detector.

For publication such matrices are frequently converted into photos and stored in photo formats (JPEG, GIF, PNG). Such photo files are good for visualization, but no longer good for data evaluation. TIFF is a special case: 8- and 24-bit TIFF files are photos, 16-bit TIFF format is a common storage format for raw scattering patterns[25], but it is not convenient for data evaluation.

Keep your precious raw data. Do not send JPEG- of GIF-encoded color images around for the purpose of "data" evaluation.

## 2.9.3 How to Utilize DI

The novice at a synchrotron facility should use the pre-evaluation options offered locally at the facility or should resort to the program *FIT2D* [39]. At least pre-evaluation operations like masking of blind areas, background correction, alignment... can be performed this way.

The real business of scattering image processing is more difficult. Because there does not exist a "point-and-click" program, the scientist must write the respective algorithms himself. In order to simplify the work, a dedicated programming system for DI should be chosen. The commercial systems IDL [40] and *pv-wave* [37] offer the key features of DI. Before choosing IDL check if it is now possible to easily write algorithms that work on matrices of varying size. Moreover, a library function for multidimensional extrapolation of data like the radial basis function [36] algorithm of *pv-wave* (RADBE) is essential. If the license fees are a problem, the free ImageJ [41] may be a solution that avoids to start from level-zero programming.

## 2.9.4 Concepts of DI that Ease the Analysis of Scattering Images

### 2.9.4.1 The Paradigm: Arithmetics with Matrices

In IDL and *pv-wave* a number, a vector, or a matrix can be handled in the same way. If s is a scattering image, b the parasitic background, and a the actual absorption factor of the sample, then a background correction[26] is carried out writing

```
wave> sc=s-a*b
```

This is not only simple[27], but computes much faster than the common concept of usual programming languages using two encapsulated loops.

---

[25] 16-bit-TIFF can be viewed and "processed" by image processing programs (photoshop, gimp,... ) the program discards the lower 8 bits in order to make the file an ordinary photo. Ergo: (1) Never overwrite your precious raw data by saving them from a program that was not made for scattering data analysis, (2) What you see in a photo processor is only a fraction of the scattering data.

[26] The example is valid for the most simple case: SAXS or USAXS in normal-transmission geometry.

[27] The same line of code evaluates curves, images or data structures of higher dimensionality (imagine time as an additional coordinate)

If we know that behind the beam stop the intensity is always 50 counts or less, we can discriminate the valid area of our image by defining a ROI mask (i.e., a shape function) (cf. p. 17) by simply writing

```
wave> m=s gt 50
```

which means: generate a matrix m of the same size as s. If a certain pixel in s is greater than (gt) 50, then set the respective pixel in m to 1 (in the logical meaning of "true"; white color; good data). The other pixels are set to 0 (in the boolean meaning of "false"; black color; invalid data).

The result will exhibit the next problem: even some of the valid pixels have not received enough photons to surmount the "blind"-level. The solution for this problem is application of the "closing" operator described at the end of this chapter.

### 2.9.4.2 Submatrix Ranking Operators

Submatrix ranking operators are belonging to the class of image-space operators (HABERÄCKER [42]) in contrast to Fourier-space operators.

**Practical Problems that are Solved by the Operations.**   Consider you have defined a mask and it turns out that pixels close to the edges of the blind areas did not receive the true intensity due to a *penumbra* effect. How do you peel off the penumbra region easily?

Consider you have forgotten to switch on "multi-read"[28] with your CCD detector and the raw data are full of cosmic-ray spikes. How do you remove them without spoiling the image?

**Definition: Submatrix.**   We choose a deliberate pixel from our scattering image. The pixel and its neighboring pixels are the submatrix. For the example we choose a submatrix size of $3 \times 3$ elements. There is scattering intensity in each pixel, e.g.

$$53 \ 68 \ 47$$
$$57 \ 67 \ 52 \ .$$
$$57 \ 64 \ 43$$

**Definition: Ranking.**   Now let us rank these values by sorting them in increasing order

$$\mathbf{43} \ 47 \ 52 \ 53 \ \mathbf{57} \ 57 \ 64 \ 67 \ \mathbf{68} \ .$$

### 2.9.4.3 Primitive Operators: Erode, Median, and Dilate

Based on the ranked list DI defines three primitive submatrix ranking operators. They determine, which value is put in the center of the submatrix:

---

[28] cf. Sect. 4.2.5.2

**erode:** Take the leftmost (smallest) value from the list (i.e. 43),

**median:** Take the value from the center of the list (i.e. 57),

**dilate:** Take the rightmost (biggest) value from the list (i.e. 68).

Erode applied to a mask peels off a layer from every "white" island in the mask. Dilate does (almost) the inverse: if erode has not managed to delete a white island completely, it is almost restored. The median operator reduces noise in the image without broadening the peaks. The operator is frequently addressed "median filter".

In practice, the abovementioned penumbra problem is solved by eroding the mask. Choose the size of the submatrix according to the width of the penumbra or try out by peeling-off thin layers repetitively. Then multiply the scattering image by the eroded mask and thus mark the penumbra region as "invalid".

In similar manner, spikes in the image from cosmic rays are extinguished by simple application of the median filter with a small submatrix size (3 or 5).

### 2.9.4.4 Combined Operators: Opening & Closing

Particularly useful are two operators that are combined from two primitive submatrix ranking operators.

**Opening (also: ouverture)** is erosion followed by dilation. The ouverture removes tiny "white" islands in the matrix if their area is smaller than half the area of the chosen submatrix.

**Closing (also: fermeture)** is dilation followed by erosion. The operator closes isolated "black holes".

Thus in order to fill small black holes in a mask, the closing operator will do automatically what otherwise would have to be done by hand[29].

From their definition the DI operators are easily implemented. Nevertheless, this implementation work is unnecessary if IDL or *pv-wave* are used, where the respective operators are simply picked from the rich library.

In fact, digital image processing systems have much more to offer – in image space, where the alignment and centering of scattering images is carried out with ease, but also in Fourier space where the predefined library functions are easily adapted to the needs of scattering pattern analysis on the fundament of scattering theory. Respective information is collected from textbooks on the field of DI and from the manuals of IDL or *pv-wave*.

---

[29]Probably by painting with the "mouse"

# 3 Typical Problems for Analysis by X-Ray Scattering

## 3.1 Everyday Industrial Problems

**How much of a crystallizable material X can I blend uniformly into a polymer until it starts to form crystals?** A series of blends with increasing amount of X is prepared. The samples are studied by WAXS (cf. Sect. 8.2) using *laboratory* equipment. Crystalline reflections of X are observed, as X starts to crystallize. Peak areas can be plotted *vs.* the known concentration in order to determine the saturation limit. Think of X being Ibuprofen and Y a polystyrene-*(b)*-polyisoprene copolymer, and you have an anti-rheumatism plaster.

**We cannot process batch X – is it no longer amorphous?** Get a sample from a processable batch Y and one from batch X. Is the transparency of both samples different? You have just carried out a light-scattering experiment[1]. Study each material by WAXS (cf. Sect. 8.2) using *laboratory* equipment and compare them. If X shows crystalline peaks, the assumption is confirmed. You may identify the peaks in order to confirm the crystallographic data. Imagine X and Y being natural rubber samples that have arrived from overseas. White zones are emerging from zones where lash belts have compressed the material: strain-induced crystallization.

**Is the semicrystalline polymer barrier getting porous during service?** Carry out SAXS measurements[2] from native and aged material. Check the initial shape of the curve at very small angles (in front of the long period peak) [43]. Is it systematically increasing as a function of ageing? This is most probably void scattering from pores. You may want to study void propagation through the barrier block: cut the block into thin sheets and check the amount of voids as a function of depth in the block. Imagine a barrier made from PVDF used in offshore business. Crystallinity can be tracked by means of an IDF analysis (Sects. 8.5.4, 8.7.3.3). Plasticizer content can be tracked by absolute intensity measurement (Sect. 7.10), measurement of the invariant and, finally, computation of the contrast.

---

[1] If the sample Y is transparent and sample X is not, X is most probably scattering light – from crystalline layers that are large enough to do so.

[2] Laboratory equipment is sufficient. Cf. Sect. 4.2.1.1

**Where is the cross-linking agent going in my thermoplastic elastomer?**
Carry out SAXS measurements calibrated to absolute intensity (Sect. 7.10). Watch
the fluctuation background (Sect. 8.3.1) and the width of the transition zone between
hard and soft domains (p. 124, Fig. 8.11). Compute the electron density (Sect. 2.2.1)
of the cross-linking agent and the ideal electron densities of the two phases (hard
and soft). Consider that phase separation in the material may not be perfect under
industrial processing conditions. Compute the contrast (p. 133). Imagine that you are
studying printing plates made from polystyrene-(*b*)-polybutadiene-(*b*)-polystyrene
block copolymers (SBS) containing acrylates that shall become cross-linked by UV-
light.

## 3.2 At the Front of Innovation

### 3.2.1 Web Resources

The worldwide web is the best source of up-to-date information concerning advanced
studies in the field of scattering of soft condensed matter. All synchrotron radiation
facilities are advertising scientific highlights, although the representation of soft-
condensed matter in these reports is varying. As this book is written, soft-condensed
matter is excellently represented at the ESRF (Grenoble, France)

> www.esrf.fr/UsersAndScience/Publications/Highlights

and good representations are available at the NSLS (Brookhaven, US)

> www.nsls.bnl.gov – search for "activity reports"

and at the APS (Argonne, US)

> www.aps.anl.gov/Science/Highlights

SPring-8 (Hyogo, Japan) has set up a "solution data base"

> http://www.spring8.or.jp/en/users/new_user/database/

which returns a list of related studies after selecting from a list of keywords.

### 3.2.2 Fields of Innovation

### 3.2.2.1 Visualize and Model Structure Automatically

This book describes methods for the *visualization and modeling* of structure from
scattering data. Some steps towards an *automated processing* of large amounts of
data have already been done, but must be continued. These methods are the basis of
quantitative research in the fields that are mentioned in the following.

### 3.2.2.2  Study Gradient Materials

Spatial variation of structure in a natural sample or in a technical part (e.g., gradient materials for hip-joints) can be studied by means of microfocus [44] beams (microbeams). The size of the probing X-ray beam limits the spatial resolution.

**2D Structure Gradients in Thin Sheets of Material.**  A thin sheet sliced from a sample is translated through a microbeam and SAXS or WAXS are measured as a function of the position on the slice. If the structure from each "pixel" of the sheet is anisotropic, the resulting data are incomplete in reciprocal space, but even qualitative analysis yields interesting results [45–48]. To overcome the incompleteness, the sample can be rotated in the beam and texture data can be recorded and analyzed [49, 50], or one can shift the scientific focus away from structure to, for example, the mapping of local strains of single carbon nanotubes as a function of bending [51].

**3D Nanostructure Gradients in Technical Parts.**  A solid piece of material is rotated and translated with respect to a microbeam and the projections of WAXS or SAXS are recorded. Virtual slicing of the material is carried out by means of tomographic reconstruction [24]. The results are scattering patterns originating from tiny cubes (voxels) in the material. Because of the rotation of the material, the scattering patterns exhibit fiber symmetry and are complete in reciprocal space. Quantitative analysis [52] is possible, even if the structure in the part is anisotropic. Tomographic reconstruction errors must be mastered in the future.

### 3.2.2.3  Study Thin Films

Very thin films exhibit special structure because of their confined geometry between substrate and surface. Their structure cannot be studied in a normal setup. In order to obtain enough photons on the detector, the X-ray beam must impinge on them under grazing incidence (Cf. Sects. 7.6.3.1, 7.6.3.2, 8.8). This technique is suitably combined with microbeams. Current effort is focusing both on progress of the instrumentation and on the development of adapted analysis methods.

### 3.2.2.4  Study Structure Evolution

The field of *in situ* studies of processes is one of the major applications of scattering methods. So there is continuous effort to extend the scope of applicability. Nevertheless, the user should be aware of the limiting factors.

**Limits of Time-Resolved and Simultaneous Measurements.**  Structure evolution studies are based on the ability to carry out time-resolved scattering experiments. The power of this scattering technique is a function of the minimum cycle time during which a scattering pattern with sufficient signal-to-noise ratio can be recorded. As cycle times for anisotropic 2D SAXS patterns have fallen below a value of 1 s at some[3] high-brilliance beamlines, time resolution is sufficient for many of the problems from the field of materials science, even if anisotropic materials are studied. Extra power is gained if several methods are coupled [53–55]. Urgently required is coupling of 2D WAXS and 2D SAXS for simultaneous measurements.

The data transfer rate is still a problem at low cycle time ($< 7$ s). In this case the *scattering patterns are buffered* during the experiment. Presentation and data transfer starts after the end of the experiment. This buffering implicates three restrictions: first, the buffer size limits the total duration of the recording. Second, there is no feedback during the experiment: the user can neither interrupt nor tune the experiment, e.g., reduce the exposure as the saturation limit of a detector is approached. Third – and this is really bad: most probably the experiment control is completely transferred to the detector, and *the clocking of the counters for the environmental parameters is deferred until the detector returns control* (cf. Sect. 4.3.1). For example, we intended to monitor force change during a fast process, and finally we get the correct time-resolved scattering patterns – but each pattern reporting the same force value (averaged over the whole period of the experiment).

**Examples.**    2D SAXS/WAXS experiments on highly anisotropic polymer materials during melting and crystallization can be used to visualize and understand the evolution of nanostructure [56,57]. Transformations of biopolymers in solution, e.g., virus crystallization can be studied *in situ* [58]. It is possible to study solidification mechanisms of spider silk [59], or the self-assembly of micelles on a time-scale of milliseconds [60].

---

[3]Many sources have been offering short cycle times for years – combined with a S/N-ratio that is sufficient for visual inspection only, but not for data analysis.

# 4 Experimental Overview

**Goniometer Setup.** The classical goniometer setup for the measurement of X-ray scattering has already been sketched in Fig. 2.2 on p. 11. In such a setup both the zero-dimensional detector (counter) and the sample are rotated. The points of the scattering pattern are measured one after another. Time-resolved measurement is impossible, but if a constant and high number of counts is accumulated at every point, the resulting pattern shows an excellent signal-to-noise ratio. Similar high-quality data are required if a study is aiming at quantitative analysis of modern nanostructured material.

**2D-Detector Setup.** Modern equipment is most frequently collecting scattering data from a whole region of the reciprocal space at the same time using a 1D- or a 2D-detector. In principle it resembles the traditional photographic X-ray cameras (pinhole camera, Kiessig camera). A sketch is presented in Fig. 4.1. Such a setup comprises beam monitoring devices (ionization chamber, pin-diode) and a detector. There is ample space to place an apparatus containing the sample in the X-ray beam.

**Fiber Symmetry: Equator and Meridian.** Figure 4.1 sketches a scattering experiment of a polymer sample under uniaxial load. Let us assume that the material

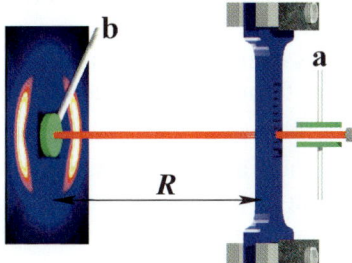

**Figure 4.1.** Typical X-ray setup with 2D detector in normal-transmission geometry. The intensity of the incident X-ray beam is measured in an ionization chamber (a). Thereafter it penetrates the sample which is subjected to some process. At a distance $R$ (cf. Table 2.1 on p. 7) behind the sample the detector is recording the scattering pattern. In its center (b) the detector is protected by a beam stop. It is equipped with a pin-diode which records the intensity of the attenuated beam

and the scattering pattern exhibit fiber symmetry. With respect to the principal axis of the sample the principal axes in the scattering pattern are associated with special names. The axis parallel to the fiber axis (i.e., in the sketch the vertical axis) is called the meridian. The axis perpendicular to the fiber axis (i.e., the horizontal axis in the pattern) is called the equator.

## 4.1 The Shape of the Primary Beam

The cross-section of the primary X-ray beam is extended and not an ideal point. This fact results in a "blurring" of the recorded scattering pattern. By keeping the cross-section tiny, modern equipment is close to the point-focus collimation approximation – because, in general, the features of the scattering patterns are relatively broad. Care must be taken, if narrow peaks like equatorial streaks (cf. p. 166) are observed and discussed. The solution is either to desmear the scattering pattern or to correct the determined structure parameters for the integral breadth of the beam profile (Sect. 9.7).

### 4.1.1 Point Focus Collimation

**Smearing.**    Because scattering is emanating from every point of the irradiated volume, the recorded scattering pattern is "smeared" by the shape of the effective cross-section of the primary beam measured in the detector plane. In terms of mathematics this smearing is accomplished by convolution (Eq. (2.17)) with the primary beam profile.

**Measure the Beam Profile.**    Deconvolution is possible if the primary beam profile has been recorded. Recording of the beam profile is readily accomplished during the adjustment of the beamline prior to the experiment as long as the beam stop has not yet been mounted. Damage to the detector is avoided[1] either by short exposure or by attenuation of the primary beam itself.

Most modern equipment is operated with small beam cross-section in order to minimize smearing effects and the need to desmear the scattering pattern.

**Desmearing.**    In practice, there are two pathways to desmear the measured image. The first is a simple result of the convolution theorem (cf. Sect. 2.7.8) which permits to carry out desmearing by means of Fourier transform, division and back-transformation (STOKES [27])

$$I(\mathbf{x}) = \mathscr{F}^{-2}\left(\mathscr{F}_2\left(I_{obs}(\mathbf{x})\right) / \mathscr{F}_2\left(W(\mathbf{x})\right)\right). \tag{4.1}$$

Here the observed scattering pattern $I_{obs}(\mathbf{x})$ is considered a digital image in 2D coordinates $\mathbf{x} = (x_1, x_3)$. $W(\mathbf{x})$ is the measured primary beam profile. Similar application

---

[1]Consult the detector manual!

of the convolution theorem is the fundament of the WARREN-AVERBACH method (Sect. 8.2.5.5, p. 107) and of RULAND's streak method (Sect. 9.7).

The second pathway is based on an iterative algorithm that was devised by VAN CITTERT [61]. The algorithm assumes that the broadening or spreading effect of the primary beam profile $W$ is approximately the same as the required narrowing effect. Its first step has become popular with the advent of digital photography where it is named "unsharp masking". After choosing the diameter of a Gaussian "point spread function", $W(\mathbf{x})$, the unsharp mask is applied and returns an "improved" image. We are satisfied if the image looks more crispy without showing nasty halos around the silhouettes. Nasty halos correspond to negative values which occur if the breadth of $W(\mathbf{x})$ was chosen too large. With our scattering patterns we hopefully have measured $W(\mathbf{x})$. So, negative values due to overdesmearing should never be observed.

The principle of the VAN CITTERT method is simple: (1) $I_1 = I_{obs}$ (2) $I_{n+1} = 2I_n - I_n \star W$ (3) stop the iteration as soon as $I_{obs} \approx I_n \star W$. This convergence criterion guarantees that the correct solution is closely approximated.

### 4.1.2 Slit Focus Collimation

#### 4.1.2.1 Common Cameras and Properties

In addition to point-focus apparatus there are scattering devices with an extremely elongated cross-section of the primary beam. Historically this geometry has been developed as a compromise between ideal collimation and insufficient scattering power. Their practical importance is decreasing as more powerful point-collimated sources become available. Kratky camera (ALEXANDER [7], p. 107-110) and Rigaku-Denki camera (BALTÁ & VONK [22], p. 83) are the most frequent representatives of slit-focus devices.

Slit-focus cameras record scattering curves. The study of anisotropic material is cumbersome. It requires large samples which can be rotated step-wise in the beam which is typically between 1 to 3 cm long.

#### 4.1.2.2 Infinite Slit Length

If, in the detector plane, the effective slit is wider than the region of the pattern in which significant intensity is observed, the approximation of an infinite slit is valid. Let the slit be infinitively long in $s_1$-direction but very narrow in $s_3$-direction then in the tangent plane approximation the recorded scattering curve

$$J(s_3) = \lceil \{I(\mathbf{s})\}_2 (s_2, s_3) \rceil_1 (s_3) \tag{4.2}$$

is the projection of the scattering intensity on the $(s_2, s_3)$-plane which is sliced in $s_3$-direction. From this relation a back-projection algorithm can be derived which permits the reconstruction of the point-collimated data from a complete set of curves recorded in a slit-focus camera.

### 4.1.2.3 A Fiber in a Slit-Focus Camera

A particularly simple case is the study of a fiber in the slit-focus camera, if the fiber is stretched out along the slit direction [31,62,63]. In this case the transversal structure according to BONART [16] (cf. Sect. 8.4.3) is directly measured, as is established by change of variables $s_1 \rightarrow s_3$, (fiber parallel to the slit) $s_2, s_3 \rightarrow s_{12}$ (fiber symmetry assumed)

$$J(s_{12}) = \lceil \{I(\mathbf{s})\}_2 (s_{12})\rceil_1 (s_1) \equiv \{I(\mathbf{s})\}_2 (s_{12}). \tag{4.3}$$

Thus, information concerning size and arrangement of domains in the cross-sectional plane of the fiber are accessible with classical laboratory equipment. Moreover, since the projection $\{I(\mathbf{s})\}_2 (s_{12})$ is complete and a normalization $\{I(\mathbf{s})\}_2 (s_{12}) \rightarrow \{I(\mathbf{s})\}_2 (s_{12})/V$ to absolute intensity units is readily established by employment of the *moving slit device*[2] without the need to resort to a secondary standard, the invariant

$$Q = 2\pi \int s_{12} \{I(\mathbf{s})\}_2 (s_{12}) /V \, ds_{12} \tag{4.4}$$

is computed after a single scan of the fiber in the slit-focus camera – provided that $V$ can be determined with sufficient accuracy.

### 4.1.3 Desmearing of Slit-Focus Data

For the case of isotropic scattering recorded with slit-focus cameras there are several desmearing options. If the slit may be considered infinite, the observed scattering intensity is

$$J(s) = \int_{-\infty}^{\infty} I\left(\sqrt{s^2 + y^2}\right) dy. \tag{4.5}$$

Juggling with projections and slices results in the GUINIER-DUMOND equation [64] (HOSEMANN [5], p. 605-607; GUINIER & FOURNET [65], p. 116-117)

$$I(s) = -\frac{1}{\pi} \int_0^{\infty} \frac{J'\left(\sqrt{s^2 + y^2}\right)}{\sqrt{s^2 + y^2}} dy. \tag{4.6}$$

for desmearing of isotropic slit-focus data. Because of the derivative in the integrand, the desmearing algorithm is quite sensitive to statistical noise on the measured data. In fact, there is no need to desmear scattering curves if the slit length is infinite and a quantitative analysis of the structure is the goal. This results from the fact that Eq. (4.5) describes an analytical projection, and if there is a method to analyze $I(s)$, it can be modified to directly act on $J(s)$, as well. A simple example is the determination of $Q$ according to Eq. (4.4). A complex example is the determination of RULAND's interface distribution function (IDF) directly from $J(s)$ [66,67].

---

[2]The moving slit device is designed to directly measure the primary beam intensity without overloading the detector. It works like a slit shutter of a photographic camera: a narrow slit is moved along the primary beam. If a sample is in the beam, the absorption of the primary beam by the sample can be directly measured.

**Table 4.1.** Performance of available point-focus setup. DORIS is an older storage ring at HASYLAB in Hamburg. The ESRF in Grenoble is an advanced synchrotron radiation source

| Setup | Flux $\left[\text{photons} \times s^{-1}\right]$ |
|---|---|
| Rotating anode, conventional optics | $1 \times 10^6$ |
| Rotating anode, Göbel mirror optics | $2 \times 10^7$ |
| Synchrotron, bending magnet (DORIS, A2) | $2 \times 10^8$ |
| Synchrotron, insertion device (ESRF, ID2) | $3 \times 10^{13}$ |

If the slit length is finite and the scattering intensity shall be desmeared, the profile $W(s_1, s_3)$ of the primary beam must be known. In order to carry out the desmearing numerically, different algorithms have been proposed, but few of these methods are able to manage the derivative problem from Eq. (4.6) properly for noisy data. One of them is the method developed by GLATTER [68].

### 4.1.4 Smearing of Point-Focus Data

In practice, sometimes it is advantageous to slit-smear point-focus data before analysis. The reason is that sometimes an expansion of point-focus data

$$I(s) = \sum_i c_i s^i$$

does not show a linear range, whereas

$$J(s) = \sum_i \tilde{c}_i s^i \approx \tilde{c}_0 + \tilde{c}_1 s$$

does. Then the constants are related to structural parameters and can easily be determined from slit-smeared data, whereas an analysis of the point-focus data may be difficult. Examples are related to the determination of density fluctuations [69] and the polydispersity in a lamellar stack [70].

## 4.2 Setup of Point-Collimation Apparatus

Since powerful X-ray sources and sophisticated beam shaping have generally become available, point-collimated setups for the study of X-ray scattering have lost their former handicap of low intensity. Today they benefit from their simple and versatile geometry. This section is devoted to an overview of modern apparatus – beginning with the source of X-radiation and ending with the detector and the data acquisition system.

The usability of the various available machines, in particular in regard to time-resolved measurements, is proportional to the flux that they are able to shine on the sample. Table 4.1 shows typical data. Modern laboratory instrumentation (rotating anode) is approaching the performance of older synchrotron light sources.

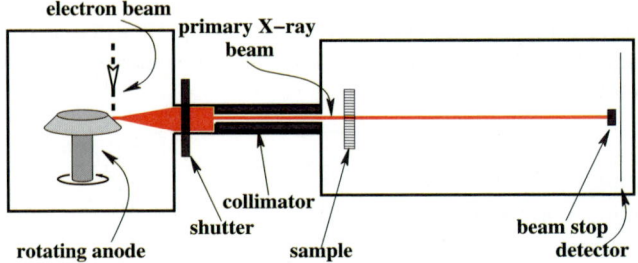

**Figure 4.2.** Sketch of a laboratory setup comprising a rotating anode, conventional beam shaping optics, and an X-ray camera with the sample in normal-transmission geometry

### 4.2.1 The Radiation Source

#### 4.2.1.1 Rotating Anode

An advancement with respect to the classical setup is an X-ray tube in which the anode is rotating. A point-focus device equipped with a rotating anode shows the same performance as a conventional system with slit-focus [71]. Figure 4.2 shows a typical laboratory setup. By rotating the anode of the X-ray tube, the power of the incident electron beam is spread on a circular ring. Thus it is possible to increase[3] the power of the tube without "burning" the anode material (Cu, Mo, ... ). High-power rotating anodes are less robust than the medium power ones. In the sketched setup the arrangement of beam, sample, and detector is called "normal-transmission geometry", because a principal axis of the sample is considered to be normal to the incident beam direction and every photon has been transmitted through the whole thickness of the sample. A different geometry is the symmetrical-reflection geometry from Fig. 2.2. Different geometries require different correction of the raw data before quantitative analysis (Sect. 7.6).

A rotating anode setup resembles a typical synchrotron beamline on a laboratory scale, and some progress concerning the optimum design of rotating setups was made by transferring sophisticated techniques for the optimization of beamline optics (PEDERSEN [72]) to rotating anode equipment.

#### 4.2.1.2 Synchrotron Radiation

**Overview.** Electrons orbiting in a magnetic field lose energy continually in the form of electromagnetic radiation (photons) emitted tangentially from the orbit. This light is called synchrotron radiation. The first dedicated synchrotron light source was the Stanford Synchrotron Radiation Laboratory (SSRL) (1977). Nowadays, many

---

[3]The maximum power of a conventional X-ray tube is 2.4 kW for broad focus (approx.. $2 \times 12$ mm focal spot size). Modern rotating anodes consume 18 kW and deliver fine focus (approx.. $0.1 \times 1$ mm focal spot size). Most important for high intensity is not the power consumption, but the product of focal spot power density and focal spot size or, more accurately, the flux on the sample measured in photons/s (cf. Sect. 7.6).

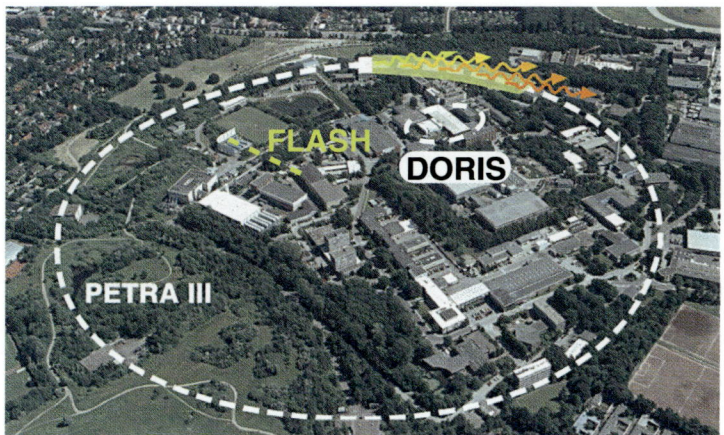

**Figure 4.3.** Three generations of X-ray light sources at DESY, Hamburg: DORIS, the small traditional synchrotron (partially enhanced by wigglers). PETRA III, the synchrotron light source of the near future. FLASH, a FEL test facility for the next generation of synchrotron light

electron synchrotrons are built exclusively for producing synchrotron radiation. The largest of these is the 8 GeV SPring-8 in Hyogo, Japan. Figure 4.3 indicates the technical progress in the field of synchrotron light sources in a sketch of the DESY site (Deutsches Elektronensynchrotron, Hamburg, Germany).

A comprehensive review on the principles of a synchrotron, its radiation and some applications in the field of soft condensed matter has been published by EL-SNER et al. [10].

**Polarization.** The central cone of the synchrotron beam from a bending magnet and, in general, the beam from insertion devices is polarized in the plane of the orbit (i.e., horizontally). Due to relativistic effects the cone of the radiation characteristics is narrow even if the beam is emitted from a bending magnet (cf. [10], p. 9-13 and Sect. 2.2.2). If necessary, polarization correction should be carried out directly at the synchrotron radiation facility by means of the locally available computer programs.

**Operating Mode of a Synchrotron Light Source.** A linear accelerator (Linac) in a synchrotron facility provides an accelerator ring with particles. In the accelerator ring the particles are repetitively accelerated by high voltages across one or several gaps while orbiting in the pipe. If the synchrotron is used for the purpose of generating synchrotron radiation, the particles are either electrons or positrons. When a desired energy is attained, the particles are transferred to a storage ring where they are allowed to cycle while the beam current is decreasing. The current is measured in units of mA and announced on monitor screens in the experimental hall. In order to keep the particles on their circular path, bending magnets are utilized. At the bending magnets synchrotron radiation is produced and emitted in tangential direction. The synchrotron radiation is polarized and has a continuous spectrum ("white light") extending from hard X-rays to the visible light. A desired wavelength is selected by diffraction in a suitable crystal, called a monochromator. The particle density in the ring is discontinuous. Several bunches are circling in the ring. Thus the synchrotron light is pulsed. The corresponding time structure is an alternating sequence of typically 100 ns light and 100 ns darkness. A different time structure ("reduced bunch mode") is required for some non-scattering experiments in which material is excited by the light pulse and the response is registered during a prolonged period of darkness.

The traditional operation mode of synchrotron light sources is a discontinuous one: particles are injected in the storage ring, the beam current is decaying exponentially, and after several hours the synchrotron radiation run is stopped for a new injection.

More and more radiation sources are switching from discontinuous mode to top-up mode. This means that the user is continuously supplied with synchrotron radiation of almost constant intensity. The loss of the electron current is either compensated continuously or in intervals of several hours (at the ESRF: 6 h).

### 4.2.1.3 XFEL: The X-Ray Free Electron Laser

According to ongoing development the most powerful and versatile X-ray light source of the next generation will no longer be realized by means of an orbit for electrons. Instead, it will be based on a long linear accelerator. The key features of this novel setup are

1. coherent X-ray light

2. tunable monochromatic X-ray light

3. typically a two-level time structure with extremely short light pulses grouped in "trains" with a train frequency of 10 Hz

Such a device should be able to accomplish the visualization of 3D nanostructure by means of X-ray holography. Moreover, the study of extremely fast processes should become possible. Since 2004, FLASH, a VUV-FEL[4] at DESY (cf. Fig. 4.3) is the first working FEL facility in the world for soft X rays. Up to the end of the year 2005 the operating parameter values of FLASH had already been decreased to a wavelength of 31.7 nm with pulse lengths down to 15 fs. During the short duration[5] of such pulses light travels a distance equivalent to the diameter of a hair.

Figure 4.4 shows the planned European XFEL Facility for hard X-rays extending from the DESY site to the East. A XFEL consists of an electron beam passing through periodic transversal magnetic fields with alternating directions. These fields cause the electrons to bend and perform a wavy motion. At each bend, very short

---

[4]V UV-FEL: vacuum ultra-violet free-electron laser

[5]Remember that short pulse means broad wavelength distribution.

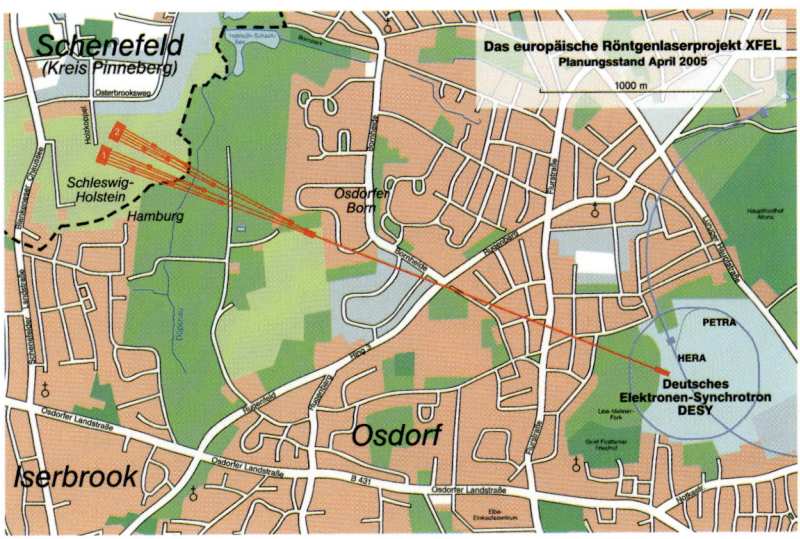

**Figure 4.4.** Proposed design of the future European XFEL Facility in Hamburg extending from DESY to laboratories in Schenefeld 4 km East. The linear tube is split into a bunch of X-ray beamlines

pulses of synchrotron radiation are emitted by the electrons. The emitted synchrotron radiation at each bend is added coherently and in this way, a pulse of short-wave nearly monochromatic radiation builds up successively. Compared with a conventional laser, a FEL can be *tuned* continuously to any wavelength, and radiation of short wavelengths can be achieved. The goal is to be able to produce monochromatic radiation down to wavelengths of 0.1 nm, which is a little bit lower than the wavelength required in the field of soft matter. The X-ray beam of an XFEL is pulsed. For the sake of versatility the time structure of the pulses is, in general, more complex than that of a synchrotron. Each of the pulses of 100 fs length contains $10^{12}$ photons[6]. 3000 of such pulses are grouped in a so-called train of 600 $\mu$s duration. The European facility is designed to operate at a train frequency of 10 Hz. It should be clear that a XFEL experiment must, in general, be synchronized with the time structure of the beam.

In 2005, the contract for the European XFEL facility had already been signed by 12 major European countries including Russia and by China. Estimated cost is 800 M€ with added 50 M€ for detector development. The facility shall be operational in 2013. There are several competitive projects around the world. The Linac Coherent Light Source (LNLS) in Stanford, USA, is under construction and shall be operational in 2009. Korea and Japan have announced respective projects. A Japanese XFEL shall be operational in 2008.

---

[6] A focused XFEL beam is not only probing the samples structure – it is as well able to excite the material – ultimately causing melting, ablation or even carbonization within picoseconds.

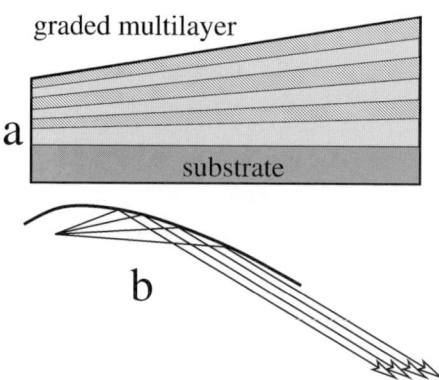

**Figure 4.5.** Beam shaping by a Göbel mirror. (a) The effective part of the mirror is a graded multilayer from materials with high contrast. (b) after *parabolic* bending the device converts divergent radiation emitted from a source point into a parallel beam

### 4.2.2 Beam Amplification by Insertion Devices

Insertion devices are placed in the electron path of a synchrotron. They increase the photon flux by several orders of magnitude. Similar to the FEL principle they operate by forcing the electrons on a wavy path. At each bend of the path synchrotron light is emitted. In contrast to the FEL device there is no coherence. Instead, the light intensity sums up to form the effective beam. Two kind of insertion devices are used. In wigglers the curvature of the electron path is high. In undulators it is relatively low.

### 4.2.3 Beam Shaping by Optical Devices

Optical devices are placed in the light path in order to shape the primary beam. Beam-position monitors, shutters, slits, monochromators, stabilizers, absorbers, and mirrors are utilized for this purpose. The effective beam shape and its flux are defined by these components. In particular, if mirrors are cooled, vibration must be avoided and thermal expansion should be compensated.

#### 4.2.3.1 The Göbel Mirror

Laboratory X-ray sources emit highly divergent radiation. With conventional optics the major part of this radiation is discarded by a slit system and a monochromator. Both components can be replaced by a Göbel mirror [73, 74]. Figure 4.5 shows its construction and application. As a result a parallel and highly monochromatic primary beam is received. Replacement of conventional incident beam optics (cf. Fig. 2.2) by a Göbel mirror increases the primary beam intensity by a factor of 10–50.

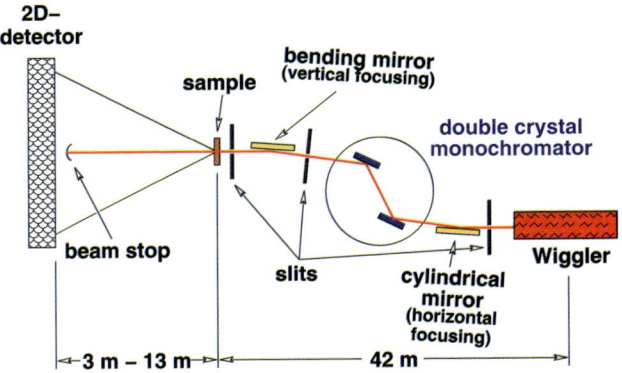

**Figure 4.6.** Optics of beamline BW4 (USAXS) at HASYLAB, Hamburg

A Göbel mirror is designed for a specific X-ray wavelength. The design concerns the choice of the two materials[7] for the multilayer, their thickness and the gradient along the mirror. Before application the graded multilayer (Fig. 4.5a) is bent and positioned (Fig. 4.5b) so that the source point of the radiation source is in the focus of the parabola.

### 4.2.3.2 Conventional Synchrotron Beamline Optics

The ordinary user who carries out scattering experiments at a synchrotron beamline will rarely adjust the optics without help. Nevertheless, during the beam time one should be able to assess the quality of the adjustment. Thus the user will most probably have to readjust some slits or to adjust the flux according to the requirements of the experiment.

Figure 4.6 presents a sketch of the BW4 beamline optics realized at the DORIS storage ring of HASYLAB, Hamburg. The beamline receives X-rays from a wiggler. Two mirrors are installed for the purpose of beam focusing. Slits define the size of the beam, and the wavelength is selected by means of a double crystal monochromator. A tilt-absorber and shutters are present, but not shown in the sketch. The beamline optics is installed in an optics hutch that is inaccessible during beam time. All components – not only the optical ones – are moved by step motors and remotely controlled via a computer program ("the motor program").

### 4.2.3.3 Microbeam Optics (Wave-Guides, X-Ray Lenses)

**Application.** Micro- and nanobeam optics are used to demagnify the cross-section of the primary beam. By means of the respective setups structure variation in inhomogeneous materials can be studied with micrometer or nanometer size resolution, respectively. For this purpose the sample is moved through the beam while

---

[7]The contrast between the materials should be big as, e.g., the one between W and Si.

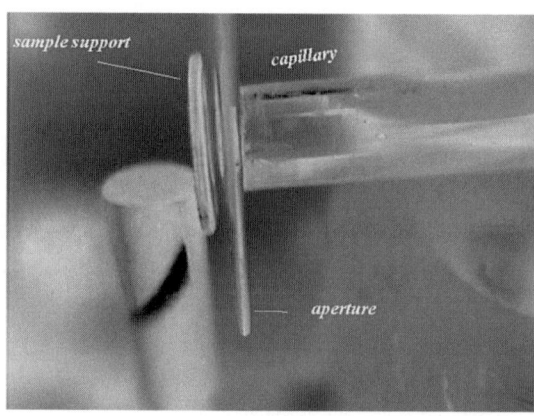

**Figure 4.7.** Microfocus beam shaping by means of a capillary used as a waveguide (ESRF, ID13) (courtesy C. RIEKEL)

scattering patterns are continuously recorded. If the sample is only translated, structure variation is studied in the plane of sheet-shaped samples (as a function of the coordinates $x$ and $y$). Additional rotation of the sample and tomographic data evaluation of the scattering patterns yields information on structure variation in space (as a function of coordinates $x$, $y$, and $z$). Tomographic measurements are time consuming. 12 hours are required for a SAXS-tomographic study of a sample with fiber symmetry [24]. Microbeam optics is advantageously coupled with other methods like grazing incidence studies [75] or microdiffraction and texture analysis [49].

**Optical Parts.**    Microbeam setups of synchrotron beamlines require special kinds of optical devices. Fresnel zone plates and glass capillaries (Fig. 4.7) used as waveguides have been known for a long time. A surprisingly effective device that can turn a synchrotron beamline into a microfocus beamline is a stack of refracting beryllium lenses [76, 77] (Fig. 4.8). The device is easily adjusted in the synchrotron beam and demagnifies its diameter by two orders of magnitude. The resulting microbeam is ready for microfocus applications.

### 4.2.3.4 Nanobeam Optics (Kirkpatrick-Baez Mirrors)

Nanobeam optics with beam diameters of several nanometers are presently developed at the ESRF. Using a Kirkpatrick-Baez optical system (cf. Fig. 4.9) beam diameters of 80 nm have been achieved. The Kirkpatrick-Baez system is made from two successively reflecting, orthogonal mirrors that are bent into *elliptical* shape by mechanical benders. The focused flux is strongly increased by deposition of a graded multilayer structure similar to that used with the *parabolic* Göbel mirror.

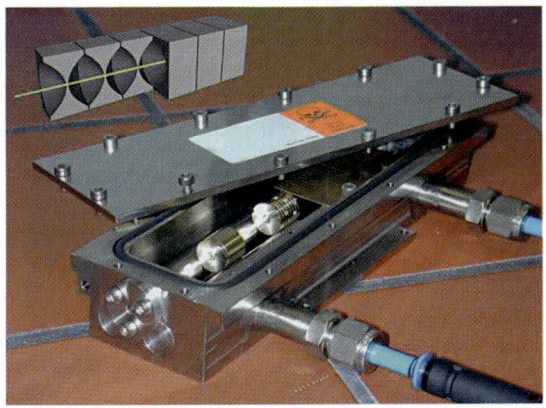

**Figure 4.8.** A stack of refracting concave beryllium lenses with parabolic profile (ESRF, France and RWTH Aachen, Germany)

**Figure 4.9.** Kirkpatrick-Baez mirrors for the generation of primary beams with nanometer size cross-section. Source: ESRF Newsletter (2005), 42(12), 14-15

### 4.2.3.5 Beam-Position Monitoring

Beam-position monitors are installed in the electron path of the synchrotron. They guarantee that the optics of each beamline "sees" the source of the synchrotron light at always the same position.

If beam position monitoring breaks down, "the beam jumps" from time to time. Such malfunction is frequently announced on the monitor screens or by the local beamline scientists. Breakdown of beam-position monitoring may be restricted to certain sectors of the ring. The user can recognize it from sudden jumps of the beam image on the detector if a semitransparent beam stop is used, from sudden changes of the slit scattering, or from sudden appearance moire patterns if the detector images are continuously accumulated into a monitor image during the experiment. In the worst case the problem will be recognized late after the experiment during data evaluation.

Result of the malfunction is that single images from a series are destroyed. Background subtraction and normalization become difficult if beam jumps have occurred during the recording of a complete set of experimental data.

### 4.2.3.6 Shutters

Shutters are integrated into the safety system of a synchrotron facility. Their operation requires training and special allowance. Each beamline has several beam shutters. The user should avoid to operate the beam shutter of the optics hutch in order not to cool down the optics. Instead, the shutter of the experimental hutch is to be operated in accordance with the interlock safety system.

### 4.2.3.7 Slits

Before any slit operation: check, write down, or save the old motor positions! Operation of slits can be useful to change the beam intensity (instead of operating absorbers). Imperfect thermal stabilization of mirrors and monochromators can be compensated by proper slit operation. Before such operation is undertaken, it should be made sure that the instrument is close to thermal equilibrium. In particular after opening the main beam shutter for the first time, it may be indicated to wait for several hours. Otherwise the operator will have to follow the thermal expansion continuously. This bears the risk to destroy the adjustment or even the detector.

Mispositioned slits result in discrete slit scattering. Slit scattering is recognized by thin and intense streaks in horizontal and/or vertical direction that extend outward from the beam stop. Take out the sample or rotate it in order to make sure that the streaks are not resulting from an interaction of the beam with the sample. Horizontal streaks can be affected by moving horizontal slit edges.

**Figure 4.10.** Beamline BW4 at HASYLAB. View from a sample chamber along the vacuum tube towards the detector at the opposite yellow wall

### 4.2.3.8 Stabilizers

Automatic adjustment of mirrors and monochromators by intensity monitors and phase-locked loops is called stabilization. If the feedback breaks down, the beam will slowly move as the temperature of the optical part is changing.

As a result slow variation of the adjustment is observed: the intensity of the primary beam will abnormally increase or decrease, the parasitic scattering background will grow, slit scattering will change (cf. Sect. 4.2.3.7). It should be clear that changes of the primary beam intensity which are paralleled by respective changes of the synchrotron current are normal.

### 4.2.3.9 Absorbers

Absorbers are found at many synchrotron beamlines. Two different principles are realized. Tilt-absorbers are operated continuously, whereas filters on a revolving disc offer step-wise attenuation of flux. Absorbers change the spectral composition of the primary beam. Thus the utilization of an absorber during scattering experiments should be avoided.

### 4.2.4 The Sample Recipient

The sample chamber contains the material that is studied in the experiment and provides for its manipulation and processing during the experiment. Figure 4.10 shows a typical sample chamber mounted at beamline BW4 of HASYLAB, Hamburg.

### 4.2.4.1 Optical Bench vs. Dance Floor

Older beamline constructions are based on an optical bench with one or two bars (cf. Fig. 4.10). This construction principle has some disadvantages

- Sample recipients have to be specially constructed or adapted for use at specific beamlines

- The construction of sample chambers is constrained by the height and the width of the bars from the optical bench

- Swapping experimental setup takes considerable time and may unnecessarily affect the optical adjustment of the beam.

**Figure 4.11.** Dance floor during construction at ANSTO near Sydney, Australia

These disadvantages are overcome by the so-called dance-floor principle which is supposed to become the major beamline construction principle of the future. Figure 4.11 shows a dance floor during the construction of the beamline hall at the ANSTO neutron-scattering facility at Lucas Heights near Sydney, Australia. The dance floor is featuring an extremely plane and hard floor surface from granite. Optical components, detectors and sample chambers are mounted on supports with a flat lower surface. While compressed air is blown into the gap between the dance floor and the area of support, components are easily moved and adjusted in the optical beam path.

### 4.2.4.2 Chambers for Sample Positioning

Some experimental techniques require the sample to be studied in very well-defined orientations and positions with respect to the X-ray beam. In the corresponding experiments the structure of the samples is, in general, not changed. A synchrotron beamline is required, because it would take too much time to record the respective data with laboratory equipment or because a special beam shape (microbeam) is essential for scanning the part with high spatial resolution.

**Complex Anisotropy** is studied in texture goniometers (p. 193) as a function of sample orientation. If the study is aiming at quantitative analysis of scattering data, the absorption correction may become an issue. Conversely, by choosing a special kind of scanning modus (e.g., symmetrical reflection SAXS; SRSAXS), the absorption correction problem can be simplified.

**Structure of Thin Films and Plane Surfaces**   is studied in a grazing-incidence recipient by means of grazing incidence (GI) methods: GISAXS, GIWAXS, or by X-ray reflectometry (XR).

**Gradient Materials**   are studied at microfocus beamlines with a spatial resolution down to 1 $\mu$m. Current development of nanobeam optics will soon allow for spatial resolutions down to 50 nm.

### 4.2.4.3  Recipients for Sample Processing

Some experiments are aiming at the study of structure evolution. In general, the studied material is isotropic or exhibits simple anisotropy (e.g., fiber symmetry). Most frequently the material is irradiated in normal-transmission geometry. A synchrotron beamline is necessary, because *in situ* recording during the materials processing is requested with a cycle time of seconds between successive snapshots (time-resolved measurements).

If a processing apparatus is constructed in such a way that the X-ray beam can irradiate the sample, it can most probably be mounted in the beamline. Suitable chambers allow for a change of sample temperature, humidity, strain, pressure, etc. Melts may be sheared during irradiation. Fibers can be spun in the beam. Several methods may be combined (SAXS, WAXS, calorimetry, light scattering) by utilization of sophisticated sample chambers.

Advanced users bring their own remote-controlled equipment, install it in the beamline and couple it to the data acquisition system of the beamline for automatic recording of environmental parameters.

In order to reduce air absorption in SAXS and USAXS setups, a vacuum tube (cf. Fig. 4.10) is mounted between sample chamber and detector.

### 4.2.5  Detectors

A variety of detectors is used in the field of X-ray scattering. In fact, the proper choice of the detector (as well as the sample thickness) is essential for good quality of the recorded data, whereas the intensity of the synchrotron radiation determines the minimum cycle time between two snapshots – if a modern CCD detector is used. Gas-filled detectors cannot be used to record high-intensity scattering patterns. Image plates need a minimum time of 2 min for read-out and erasure.

The basic principles of position-sensitive detectors are published in an early review of J. HENDRIX [78]. A review on 2D X-ray scattering of polymers including a description of detectors has been published by RUDOLF & LANDES [38]. WILSON [79] describes the detector development at EMBL, Hamburg which contributed to the commercial success of *mar research Inc.*

### 4.2.5.1  Criteria for Detector Performance

The key features concerning detector performance are

- a long linearity range (minimum 0...14000)

- a high dynamical range (minimum 14 bit = 0...16383)

- built-in correction for detector response

- automatic correction for cosmic-ray spikes

- fast read-out (not slower than 1 s)

Some of these issues are features of the detector software that must be switched on by the user.

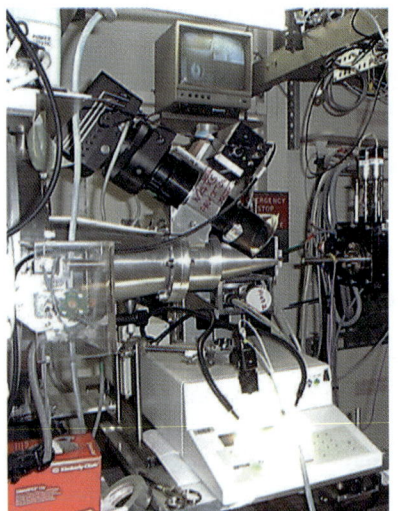

**Figure 4.12.** WAXS and USAXS detectors combined at beamline ID2, ESRF

A challenging task is the recording of anisotropic data in different ranges of scattering with different detectors. In this case the placement of the detectors becomes the critical point. An optimum solution would be to echelon ring-shaped WAXS and SAXS detectors with a disc-shaped USAXS detector placed at a long distance. A versatile solution using available technology has been realized at beamline ID02 of the ESRF. Figure 4.12 shows the black WAXS detector mounted in an inclined orientation close to the sample position. The USAXS detector is not visible in the picture. It is mounted at the far end of the silver vacuum tube running to the left. In the presented setup the sample is kept in a heating stage that can be moved into the beam.

### 4.2.5.2 CCD Detectors

Not only in the field of digital photography, but also in the field of scattering CCD detectors have become very popular. In particular the commercial[8] mar CCD detectors have boosted the research in the field of soft matter. They feature robust remote operation, short read-out time, high spatial resolution, automatic internal compensation of non-uniform detector response and cosmic-ray spike removal in a so-called "multi-read" modus. Very important is the fact that saturation is only a matter of accumulated counts, not a matter of scattering intensity (counts / second). Thus saturation can be avoided by reducing the exposure time. Moreover, if saturation occurs it is limited to a distinct region of the scattering image but does not spoil the whole

---

[8]mar CCD is a trademark of *mar research Inc.* (www.marresearch.com). The president of mar research is Jules Hendrix, a detector specialist formerly employed at the EMBL outstation of HASYLAB, Hamburg.

**Figure 4.13.** mar® 185 CCD–detector in operation at beamline BW4, HASYLAB at the rear end of the vacuum tube

image, as it is the fact with gas-filled detectors. Figure 4.13 shows a mar CCD 165 detector in operation at beamline A2 of HASYLAB, Hamburg. The diameter of its sensitive area is 165 mm. This area is subdivided into 4 CCD chips. The detector's target CCD chips are cooled[9] in order to reduce the dark current. The digital image size is 2048 × 2048 pixels with a depth (dynamical range) of 16 bits. The image is stored in 16-bit TIFF format. There is a linear relation between scattering intensity and pixel value. A constant readout background is found in each pixel (typically 10 counts). Readout: 2.7 s for the highest resolution, 0.7 s if the resolution is reduced to 1024 × 1024 pixels. Energy discrimination is impossible, i.e., the detector is sensitive to fluorescence photons as well.

### 4.2.5.3 Image Plates

Image plates are the modern equivalent to photographic film. Readout after exposure (here it is called "scanning") is something like a photographic development process: typically it takes about 3 min. Thereafter another 3 min are required to erase the image plate. Image plates are sensitive both to mechanical damage and daylight. After erasure they slowly accumulate a "fog"-background. The spatial sensitivity of image plates to X-rays, i.e., their "detector response" is non-uniform. Image plates are subjected to ageing: their detector response changes with time. Advantageous is their high spatial resolution and their robustness in regard to overexposure. Holes can be drilled into image plates in order to let the SAXS through to a second detector while the WAXS is recorded on the plate (simultaneous measurement). The ordinary image plate cannot be operated remotely. The image plate scanner returns 16-bit

---

[9]Keep an eye on the cooler. Switch it on upon restart of the detector.

TIFF data, the so-called "GEL-format". The GEL-format is nonlinear: instead of the intensity a number that is proportional to the square root of the intensity is stored in the file.

A user-friendly environment for an image plate has been invented by *mar research Inc.* In the mar345 detector an image plate, a scanner and an eraser are integrated in a single device. The image plate is a circular disc and the scanner-eraser unit is operating after exposure similar to the arm of a classical turn-table gramophone. In this environment the image-plate detector can be operated remotely and the detector characteristics of the image plate can be compensated by the read-out electronics.

Image plates use stimulated luminescence from storage phosphor materials. The commercially available plates are composed of extremely fine crystals of $BaFBrEu^{2+}$. X-rays excite an electron of $Eu^{2+}$ into the conduction band, where it is trapped in an F-center of the barium halide with a subsequent oxidation of $Eu^{2+}$ to $Eu^{3+}$. By exposing the $BaFBrEu^-$ complex to light from a HeNe laser the electrons are liberated with the emission of a photon at 390 nm [38].

### 4.2.5.4 Gas-Filled Detectors

Gas-filled detectors are the classical X-ray detectors. The main advantages are

- the possibility to carry out energy discrimination and thus only to count elastically scattered photons

- the close relation between a count in the electronics and an incident photon ("single photon counters")

The major disadvantages are

- constant *dead time* after each pulse that is counted in the gas volume. This fact leads to an *integral* intensity limit for the whole detector – typically not higher than $10^5 \, s^{-1}$.

- Non-uniform detector response – wire-structure visible in the raw data.

These features made gas-filled detectors the limiting factor for effective use of synchrotron beamlines. Nevertheless, they are still well-suited for laboratory equipment[10]. Gas-filled detectors are classified by their dimensionality.

**0D Detectors**   are the classical proportional counters that are used in laboratory goniometers for decades. Because every point of reciprocal space is measured with the same cell, the detector response is uniform by definition.

---

[10]If the progress of laboratory X-ray sources continues, gas-filled detectors will probably be replaced by image plates with automatic read-out.

**Figure 4.14.** A one-dimensional bent detector (INEL CPS120) can record a complete WAXS curve simultaneously (source: www.imp.cnrs.fr/ESCA/drx.htm)

**1D Position Sensitive Detectors.** Position sensitivity is accomplished by a so-called delay line. For every pulse[11] arriving at the wire the time is measured that it needs to travel to each of the two ends of the wire. Thus the position of the incident photon along the wire can be computed from the time difference, i.e., the delay. Bent high-resolution 1D position sensitive detectors (cf. Fig. 4.14) are advantageously used in laboratory equipment for the recording of WAXS curves.

**2D Position Sensitive Detectors** are multi-wire electrical-field detectors. The principal limitation of the total counting rate reduces the applicability at a synchrotron beamline in particular for 2D detectors. But even strong, narrow peaks pose a problem, because the whole image is distorted as soon as local saturation occurs. The detector response is changing, because the wires are worn out by use.

### 4.2.5.5 Other X-Ray Detectors

Several other principles have been used to build X-ray detectors. For instance, 1D detectors have been realized by diode arrays. 2D detectors have been realized by conversion of X-rays to visible light, photon amplification, and a television camera (VIDICON). CCD detectors have outperformed both diode arrays and the VIDICON.

---

[11] Every pulse is an electron avalanche generated by ionization of the counter gas as a result of the interaction with the incoming photon

### 4.2.5.6 Detector Operation Mode: Binning

In practice, the user selects the spatial resolution combined with the readout time of a modern 2D detector. The lower resolutions are realized by binning[12] of pixels on the detector. Typical ranges are $1024 \times 1024$ pixels with 0.7 s readout time and $2048 \times 2048$ pixels with 7 s readout time. A low resolution should only be selected if the high-resolution readout time is too long for the experiment.

The reason is truncation of dynamics in low-resolution modes: in a 14-bit-detector each pixel in low-resolution mode can contain 0 to 16382 counts. With the next photon an arithmetic overflow will occur and the pixel is saturated. In high-resolution mode the same area of the detector is represented by 4 pixels, and if the intensity is evenly distributed it takes 4 times longer before the pixels will be saturated. If the high resolution is not required and the cycle time is 30 s or longer, it is good practice to store away the big files on a spacious USB hard-disk and afterwards to bin the data.

### 4.2.6 Experiment Monitors

Data analysis is based on the proper tracking of the key environmental parameters of the experiment. Apart from the obvious parameters (e.g., the elongation in a straining experiment) there are some essential parameters that are related to the scattering apparatus itself (e.g., the intensity of the incident beam).

### 4.2.6.1 Monitoring, Journaling, Control

**Automatic Monitoring.** The success of the experiment is strongly related to the ability of the data acquisition system to automatically record the relevant parameters during the experiment. Monitoring basic parameters by paper and pencil is only the last resort.

**Human Journaling.** One of the team members should observe the running experiment and the data acquisition system and write down the observations of the team in a journal file on a laptop computer.

**Remote Controlling.** X-rays are harmful to humans. Nobody is allowed to enter the experimental hutch during experiments with X-rays. If observation and control is important, the experiment is observed by a system of TV-cameras and TV-monitors and controlled remotely by a second scientist.

TV-cameras should be tiny. Otherwise they cannot be placed almost parallel to the optical axis of the X-ray beam. The focus should be adjustable in a wide range.

---

[12]Binning means that the intensities from four physical cells in a square are added into a single virtual pixel. By this operation the edge length of the scattering pattern is halved. Because the detector hardware only permits to store data in a certain width (usually not more than 16 bits), overflow may occur upon binning "on the detector". As a result, the corresponding areas are marked as saturated – only for the reason of limited arithmetical capabilities of the detector hardware.

Suitable "camera modules" which can picture an area down to 7 mm × 10 mm on full screen are available from electronics suppliers at a price of less than 20 €.

### 4.2.6.2 Beam Intensity Monitoring

The intensity of the X-ray beam is measured by ionization chambers or pin-diodes[13]. Pin-diodes can only be operated in the beam stop. The variation of the beam intensity during the experiment should be measured both before and after the sample. If the beam intensity monitors are set up properly, the absorption of the primary beam by the sample can be computed for each scattering pattern. The placement of the "*first ionization chamber*" in or after the X-ray guide tube to the sample is uncritical.

The design and placement of the *second* beam intensity monitor demands more attention. The definition of X-ray absorption does not discriminate between primary beam, USAXS and SAXS. So the second beam intensity monitor should guide primary beam, USAXS and SAXS through its volume, whereas the WAXS should pass outside the monitor. The optimum setup for SAXS and USAXS measurements is a narrow *ionization chamber directly behind the sample*. For WAXS measurement a *pin-diode in the beam stop* is a good solution for WAXS. For USAXS and SAXS it may be acceptable, as long as the relevant part of the primary beam is caught, the optical system is in thermal equilibrium and the synchrotron beam does not jump (cf. Sect. 4.2.3.5).

## 4.3 Data Acquisition, Experiment Control and Its Principles

In order to operate a complex X-ray device professionally, some understanding on the underlying principle of data acquisition and experiment control is helpful. Figure 4.15 shows the monitor rack of a synchrotron beamline. This monitor rack has 8 monitor channels to report 8 different signals which describe environmental parameters of the experiment. In case of a malfunction, the displayed numbers may be written down using paper and pencil. Additionally, the rack is used for the association of the signals to the monitor channels of the data acquisition system by plugging cables with LEMO®[14]FFA-plugs into respective sockets.

### 4.3.1 Voltage-to-Frequency Conversion (VFC)

The basic measurement principle of an X-ray device is counting. We open a beam shutter, count discrete pulses, and close the shutter again. The shutter is mimicked by electronics which opens a gate to a counter for a certain period of time. After the

---

[13] A pin-diode has three layers: p-doted layer, i: intrinsic interaction layer, n-doted layer. The outer layers provide the electrical field. In the inner layer photons generate electron-hole-pairs which result in a current, although the diode is operated in reverse-biasing mode.

[14] http://www.lemo.com after the Swiss engineer LÉON MOUTTET – A company manufacturing precision connectors. The user will find these connectors at many beamlines around the world and will have to provide his signal cable with a respective plug.

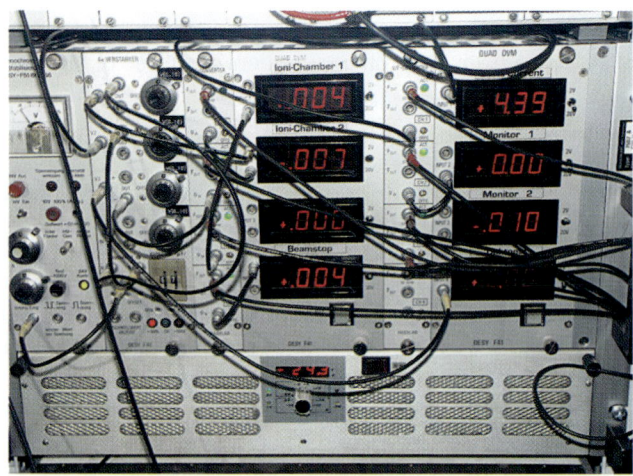

**Figure 4.15.** Digital display and wire connections related to the 8 monitor channels which can be managed by the environmental data acquisition system of beamline BW4, HASYLAB, Hamburg

gate is closed the value of the counter is read out. It is the paradigm of X-ray data acquisition that *every signal is converted to frequency in order to make it "countable"*. Thus all environmental parameters are "counted" during the exposure of a scattering pattern. This has several consequences

- Monitored parameters are related to channels of a multi-channel counter. The number of counts found in the counter channels after each exposure will be reported in a monitor file [15] together with a time stamp.

- Each parameter value reported in the monitor file must be divided by the actual exposure time in order to yield a reasonable value. For this purpose one of the channels should be reserved and count the "heartbeats" of an oscillator quartz.

- If a signal is varying during exposure, the value obtained after division is exactly what we need: the time-average over the exposure interval.

- It is good practice to calibrate the monitor channels with respect to the parameters[16] before starting the experiments.

- Check the output in the monitor file and the functionality of the counters[17] several times during the beam time.

---

[15] This file may be a part of the scattering pattern file or a file of its own. For the sake of easy viewing the monitor data are, in general, human-readable text ("ASCII–header" or" ASCII –file").

[16] Temperature in degrees; moved distance in length units; etc.

[17] Counters should run while the shutter is open and should be quiet during data read-out. Otherwise a gate flip-flop has changed its state unintentionally and the data acquisition system must be rebooted.

- There is a shortest reasonable exposure time that is related to the frequency of the heartbeat quartz and the adjustment of the voltage-to-frequency converters. Below the reasonable exposure interval quantization errors become a problem and the "measured" value will be chosen from a small number of possible steps.

- If the external gate is no longer clocking the counters because the shutter control has been transferred to the internal detector software, no *variation* of the environmental parameters will be recorded any more (For example, if you ask your detector to accumulate 200 scattering patterns with a very fast cycle time "in its buffer", there is only one gate-opening for the start of the series and one gate-closing at its end. Thus, all your 200 images will report the same environmental data averaged over the whole series of 200 images).

- At the beginning of a beam time the users should write down which monitor channel is connected to which signal.

- Users provide their environmental parameters as an electrical voltage (typically in the range of $0 \ldots 10$ V) and connect a cable to an input of a VFC of a free monitor channel.

### 4.3.2 Unix and the Communication Among Acquisition Modules

The components of the data acquisition system are communicating with each other via Internet protocol (TCP/IP). For instance, detectors are listening at Internet ports for commands to change their mode of operation and broadcast acknowledgments which are intercepted by the experiment control program. This is, in general, accomplished by a network of computers working under various Unix variants. It is thus very helpful, if the user has some experience in Unix. This starts from the question of how to check the raw data files for completeness and ends with complex scripts to simplify the control of the experiment[18].

---

[18] If, e.g., during a running experiment the danger of detector saturation is observed, the experienced user may broadcast a command to the detector (by writing into a pipe): change the exposure but keep the cycle time constant.

# 5 Acquisition of Synchrotron Beamtime

Regular beamtime at a synchrotron radiation source is allocated by a review panel on the basis of a written proposal. The main part of this short chapter is a guide to the novice with hints to the writing of a regular proposal. Besides this regular access there are, in general less laborious ways to put a sample in the synchrotron beam.

## 5.1 Test Measurements

The advantage of test measurements is easy access. The disadvantage is the fact that there is no funding of travel expenses and accommodation.

At some synchrotrons the novice can *officially ask for test-measurement beamtime*. Such beamtime will not be longer than one day and will be appended to beamtime that uses the same setup as the one required for the test.

Another possibility is to *ask an experienced colleague with a granted proposal* to join his beamtime with some samples that fit into his project. In this case there may even be a chance to be funded within the frame of this collaboration.

The beamline scientist at a synchrotron beamline has some *in-house beamtime* of his own. The test-user may convince the staff scientist to study a test sample during this beamtime. It is possible that the user simply sends the sample, but it is better to join in the measurement.

After having performed such tests the new user should be able to assess whether it appears reasonable to study the scientific problem at a synchrotron beamline. Sometimes one will simply be able to use devices (furnace, extensometer, sample recipient) provided at the beamline. Sometimes the researcher will have to adapt some own devices to fit in the beamline, to control it remotely and to record its output signals together with the scattering patterns. Sometimes special equipment will have to be constructed.

## 5.2 Support or Collaboration

Early decision should be made whether the user only requests support from the beamline staff, or whether a scientific collaboration is offered.

**Supported Beamtime.** In a supported beamtime, the beamline scientists and their engineers will adjust the beam and help with the setup of provided devices. They will instruct the user. Thereafter they will be on stand-by for the case of problems. Finally the staff will provide means to transfer the scattering data. It is the user's duty to report the results of the study to the synchrotron radiation facility (forms are available for that purpose). Publications based on such study must give proper reference to the synchrotron radiation facility.

**Acknowledgments.**   Regrettably it must be said that only a minority of the scientific users pays tribute to the beamline staff. Even if the staff is not collaborating in the project, an acknowledgment should be the minimum courtesy in any publication. An example for such an acknowledgment is *"We acknowledge <facility>, <place>, for provision of the synchrotron radiation facilities at beamline <name> in the frame of project <number>. In particular the support of <scientist> and the beamline engineer <name> is greatly appreciated."*. Variants can be found on the web-pages of synchrotron radiation facilities that deal with this problem.

**Collaboration with Beamline Staff.**   If the beamline staff accepts scientific collaboration with the user, the beamline scientists will actively participate in the experiment and the engineers will help with the adaption of special devices. Such active cooperation should be awarded co-authoring the resulting papers, in particular if the colleagues have participated in the data evaluation and in the discussion of the manuscripts.

Disrespect of etiquette has little impact on the review panel. For the panel it is important that the results are properly reported and published. Users who do not report will not receive a follow-up beamtime in a similar project. Users who do not publish for years will only receive beamtime if no productive user is competing.

## 5.3 A Guide to Proposal Writing

This section describes the situation at the ESRF up to the year 2003.

**How a Review Panel Works.**   Be aware of the fact that the members of the review panel have to decide on 120 proposals every 6 months during one week. The panel has 7 members. After the members have received the proposals they read every proposal and rank it individually during three days. For every proposal two members are elected speakers. When the panel meets for two days, the individual rankings have been collected, and the secretary has prepared a list giving the average ranking and the standard deviation. Promising but poorly written proposals are characterized by a wide standard deviation. Every proposal is introduced by the speakers, who try to give information that might be missing in the paperwork. In this way the panel tries to find a fair ranking for those proposals with a wide standard deviation, whereas the proposals with a narrow error bar are not discussed in detail. Finally, beamtime is allocated in the order of the ranking.

**What the Panel Member Does not Need to Know.**   It is generally conceded that your scientific project is very important. A lengthy introduction intended to convince the panel by giving information that is not closely related to the experiment is exhausting, in particular if it fills half of the proposal form. The panel member gets annoyed, if by this procedure important information has been squeezed out. The panel member who is the speaker will probably have to retrieve the missing information from the Internet. The panel member becomes more annoyed if this happens for several proposals in succession. The panel member is happy if he reads a clear and concise proposal.

**What the Panel Member Must Know.**   Do not write more than 10 lines on the *impact* of the expected results. Only one of the speakers is an expert in your field. Address the interested lay-person! Are the expected results of general public interest so that they can advertise the research facility in the public press? This is a strong argument to the panel.

Document or explain the *feasibility* of your experiment. Explain, why the experiment must be carried out at a special beamline. For overbooked facilities explain, why the experiment must be performed there and cannot performed at a low power synchrotron[1]. If there are several alternative beamlines where the experiment can be performed: show the alternatives. The resulting flexibility for the panel increases the chance to become allocated.

Sketch the *setup* of your experiment (sample-to-detector distance[2], requested detector(s), special sample environment requested from the facility or brought with you), and your *experiment plan* (how many samples? What parameters are varied?) and deduce from it the number of *requested shifts or days*.[3]

Show your *expertise* or document that you are collaborating with an expert. In particular, indicate how you intend to evaluate the collected data and reference relevant literature.

---

[1]Present results from laboratory sources, low power sources or previous experiments (also from others). Such data are strong arguments.

[2]The choice of the sample-to-detector distance, $R$, is a problem of SAXS. Let $L$ be the expected long period the material to be studied and $D$ be the diameter of the 2D detector, then $R \approx LD/(9\lambda)$ is a good first guess.

[3]On an overbooked beamline try to devise an experiment plan that requires 3 to 4 days. In this case the beamtime is long enough for the staff: they do not have to change the setup more than twice in a week. On the other hand, the request is flexible enough for allocation by the review panel.

# 6 It's Beamtime, Phil[1]: A Guide to Collect a Complete Set of Data

When the experiment has been set up at the beamline or the rotating anode, it is important to collect a complete set of data for later evaluation or the experiment will have to be repeated. Do not expect to be satisfied after the first time, because the collection of complete data is a matter of experience. The intention of this chapter is to reduce the number of repetitions.

## 6.1 Be Organized

In advance of the beamtime discuss the experiment with an expert. At a synchrotron beamline cooperate with the local beamline staff.

It is reasonable to have a blotter and a pencil at hand. Label the blotter with the experiment and date. **Write down on paper all information concerning smooth and complete conduct of the experiment.** Paper and pencil are handy during the briefing by the local staff when you are talking and walking around on the site. After the beamtime copy respective information from your journal file to the blotter. Continue to do so during data evaluation. **Before the next beamtime consult the old experiment blotter for preparation**. Your own final guide will probably contain several of the issues that are addressed in the following sections of this chapter.

## 6.2 Very Important: Data File Check

After you have collected the first scattering patterns check that the data files have arrived in the expected directories. Check the size of the files. Open the ASCII files or the ASCII headers in a text editor and check that the environmental data have arrived in the files. Vary environmental parameters in test measurements and check that the values in the ASCII files vary accordingly. If possible, calibrate environmental parameters (e.g., sample temperature, straining force, cross-bar position). Ask the beamline staff to demonstrate what they tell you. Double check! Otherwise your effort may be wasted.

---

[1] "Don't drive angry!" – Bill Murray in the movie *Groundhog Day*

## 6.3 Never Store Test Snapshots from Detector Memory

From time to time you will certainly perform test snapshots[2] that are only accumulated in the memory of the detector before you start the real business. If everything looks fine – do not store the test image from the detector memory. Instead **repeat the snapshot using the command that directly stores the data in a file**. Storage from memory will most probably not dump the correct environmental data to the data files – and if afterwards you want to use the data as, e.g., the machine background and you find out that the exposure and the primary beam intensity have not been stored, you may have a problem.

## 6.4 To Be Collected Before the First Experiment

In this section the static parameters of the experimental setup are collected. It is good practice to do so after each re-adjustment of the setup. If some of the data addressed here have been forgotten to be collected there is some chance that the beamline scientist has done so during adjustment.

- X-ray wavelength

- sample-to-detector distance $R$

- technical description of the detector(s) used

- detector response of image plates and gas-filled detectors

- size of each pixel cell on the detector

- allocation of monitor channels to environmental parameters

- frequency of the heartbeat monitor channel

- calibration of monitor channels for which absolute data values are required (e.g., sample temperature, sample elongation)

- a measured primary beam profile or the integral breadths of the primary beam on the detector in horizontal and vertical directions

- in particular in studies of porous or fractal materials, assessment of multiple scattering (cf. Sect. 7.2) should be carried out

The parameter values and patterns determined in this section are reasonably linked into adapted versions of data pre-evaluation procedures in order to permit automated pre-processing of complete series of scattering patterns. Novices at synchrotron beamlines will resort to on-site procedures and take home pre-processed data.

---

[2]Frequent reasons are that you (1) want to test what exposure will be needed, (2) have to check the slit scattering of the apparatus, (3) want to know if the beam comes through, (4) do not know if the sample is in the beam

### 6.4.1  Measurement of the Sample-Detector Distance

For SAXS and USAXS experiments it is sufficient to measure the distance $R$ by hand or by a laser distance-meter. For MAXS and WAXS, calibration by means of samples with known sharp reflection are required. Such calibration samples are available at synchrotron beamlines. Different materials are used for application in different angular regions (isotropic crystalline materials for WAXS, Ag-behenate for MAXS, diverse biological samples[3] for SAXS and USAXS). Calibration sheets are frequently published on the home pages of synchrotron beamlines.

### 6.4.2  Measurement of the Detector Response

Some commercial detectors come with built-in procedures and software that automatically corrects for a flat detector response of every pixel[4]. For other detectors the necessary correction has to be carried out by the user.

The detector response is measured by placing a radioactive iron-source ($^{55}$Fe) in front of the detector and accumulating for several hours. If the source is placed at a short distance of the detector, the varying distance of each pixel from the source must be considered. Actual pre-evaluated[5] detector response images should be available from the local staff. Detector response is corrected by multiplying the pre-evaluated detector response image and the raw data accumulated during the experiment.

### 6.4.3  Measurement of the Primary Beam Profile

The primary beam profile is reasonably measured during adjustment of the optics just before the beam stop is inserted. If overexposure of the detector can be avoided by choosing a short exposure interval this method is to be favored. Instead, attenuation of the primary beam by an absorber must be considered.

## 6.5  To Be Collected for Each New Run

If the synchrotron is operated in discontinuous mode, the storage ring will be refilled two or three times every day. The interval between two consecutive refills is called a synchrotron radiation run. The parasitic scattering (*machine background*) should at least be recorded once within each synchrotron radiation run. You might consider

---

[3] Rattail tendon, ox-eye cornea . . .

[4] Such detectors reveal their built-in intelligence if the user interrupts a measurement. Thereafter the detector response memory may contain nonsensical data, and the following measurement may be corrupted.

[5] The intensity measured during response calibration must be corrected for intensity loss due to the varying distance of each pixel from the iron source. Thereafter the invalid pixels must be masked. For the valid pixels the reciprocal intensity is computed and finally normalized in such a way that the average intensity of all the valid pixels is 1 in order not to introduce global intensity change by the following response corrections.

to measure a background before every experiment, because there is some probability that the beam position jumps by a few pixels during refill of the storage ring. Moreover, temperature variation may cause slow variation of the adjustment.

Background is measured without the sample – but including the sample holder and all kind of "wrapping" that does not belong to the material of interest (e.g., aluminum foil, empty cuvette, or cuvette with solvent, respectively).

The exposure time of the background should be prolonged by a factor of about 5 compared to the exposure of the sample. Thus a better signal-to-noise ratio is obtained, and this decreases the risk to obtain negative intensities upon background correction. If prolonged exposure is not possible the risk of negative intensities can later be minimized during data evaluation by median-filtering (Sect. 2.9) the background before using it. A submatrix width of 5 works fine and it may even be increased if there is no significant slit scattering in the background pattern.

## 6.6 Adjustments with Each Experiment

**Sample-to-Detector Distance** Perform test exposures and check that no relevant peak is "cut" on the detector image – outside the peaks there should be a considerable amount of the diffuse tail of the scattering recorded. If the distance is too short only for some of the samples one may consider to first study other samples and then to re-adjust the beamline. Assess the extra effort and anticipate that the beamline scientist might not be pleased. It is better practice to plan ahead and to test some materials for future beamtimes. Estimate now the distance needed next time.

**Exposure Time** Perform some test exposures and choose an exposure time so that the maximum scattering originating from the sample results in about 10,000 counts in some of the relevant pixels. Is the maximum lower than 2000 counts, a quantitative analysis will be very difficult.

**Sample Thickness** If possible choose a sample thickness close to the optimum thickness (Sect. 7.6). Proper background correction will be difficult for extremely thin samples. It is better to shorten the exposure or even to attenuate the primary beam than to decrease the sample thickness far below the optimum thickness.

**Sample Orientation** If possible orient the sample in such a way that the beamstop holder does not cut through an important region (peak). If you expect that the sample exhibits fiber symmetry, check it: rotating the sample about the assumed fiber axis and take some patterns.

## 6.7 Collect Good Data

Spend time on data accumulation if you plan to evaluate the data. Exposure of one WAXS curve from a modern thin-film sample that is good enough for line profile analysis takes several hours at a laboratory X-ray source – even if a rotating anode, Göbel mirror, and a bent position sensitive detector is operated. Immediately after the experiment cut curves from the data and check the S/N–ratio.

## 6.8  To Be Collected with Each Scattering Pattern

The parameters of this category should be recorded automatically using monitor channels of the data acquisition system (cf. Sect. 4.3). The value of the experiment is strongly resting on these data. In case of a malfunction or unavailability of the monitor channel module of the data acquisition system at least the most important parameter must be collected using paper and pencil. These parameters are the main process parameter[6], the exposure times, and the readings of the ionization chambers.

---

[6]For instance, the main process parameter in a heating/cooling experiment is the temperature.

# 7 Pre-evaluation of Scattering Data

Unfortunately the standardization of evaluation procedures is still low in the field of scattering as compared with, e.g., crystallography. Thus, whenever scattering data shall be evaluated the user has to learn how to write computer programs. There is no possibility to circumvent learning this lesson by using one of the standard data evaluation programs like OTOKO[1], TOPAS[2], or *FIT2D*[3], because these programs do not provide the user with much built-in intelligence. Instead, they represent a library of algorithmic tools which have to be parameterized and chained up reasonably by the user. In fact, this chaining is interactive programming, and the resulting chain of algorithms is a computer script. It is adapted to the specific beamline setup and should be capable to semi-automatically pre-evaluate[4] a complete sequence of frames, i.e., a stream of X-ray scattering patterns.

If one of the invoked algorithms fails, it is most probably fed with nonsensical data. For instance, if an algorithm needs to extrapolate intensity to zero scattering angle it is nonsensical to provide it with a scattering curve that starts with a sequence of invalid intensities from behind the beam stop. The algorithm will most probably break down, if these points are not removed or marked as invalid. If a program breakdown has occurred, it often helps to imagine how the scientist himself would solve the task step-by-step using paper and pencil.

The data evaluation process is divided into pre-evaluation and structure evaluation. The pre-evaluation follows quite simple rules, which are a function of the setup geometry and the data quality only. During this first stage the scientist observes the features of the scattering patterns. From these features he learns how to tackle the structure evaluation for his specific case.

Full automated data pre-evaluation can rarely be achieved. In particular, centering and alignment of each raw pattern should at least be controlled by the user. On the other hand, even semi-automated pre-evaluation can

---

[1] http://www.embl-hamburg.de/ExternalInfo/Research/Ncs/otoko.html

[2] http://www.chemie.uni-hamburg.de/tmc/stribeck/dl/

[3] http://www.esrf.fr/computing/scientific/FIT2D/

[4] Pre-evaluation can be made to work automatically under certain conditions: (1) each frame carries an individual base name through all steps of data evaluation (e.g., "myexperiment_frame003") which can be addressed by the script (e.g., by referencing "image.name") – (2) the parameters of the detector image file (e.g., pixel size, width, height, exposure time stamp, exposure time) can be referenced by the script, can be attached to the image and carried along with it through all steps of evaluation for deliberate use – (3) the monitor channel parameter values (cf. Sect. 4.3) are accessible in a similar manner and attachable to general purpose "experimental parameter variables" of the image.

be obstructed by poor design of the X-ray beamline or by a poor experimental setup [80]. The pre-evaluation procedure must be able to read every parameter[5] it needs from a file which was generated during the experiment.

## 7.1 Reading the Scattering Data Files

There is no standard data format for scattering data. The vast list of supported file formats in the reference manual of *FIT2D* clearly shows the problem.

Using a readily available computer program has the advantage that it offers predefined algorithmic tools. The disadvantage is that the toolbox is rarely complete. So the wish to read the data files into a spreadsheet program, into Origin or another program of the user's choice, is quite common. If the experience[6] in programming is low, the to-date best advice is to use OTOKO or *FIT2D* for reading the raw data, to perform pre-evaluation, and, finally to output the data in a suitable format for storage, presentation or further treatment by a different program.

If the user has a favorite evaluation program that he knows how to operate, he may use *FIT2D* as a converter program only – or he may directly resort to a stand-alone format converter program. Converters which output at least human-readable ASCII format are frequently available at synchrotron radiation facilities – and ASCII file import should be supported by any reasonable program.

## 7.2 Assessment of SAXS Multiple Scattering

Multiple scattering means that a photon is at least scattered twice inside the sample. The phenomenon changes the measured small-angle scattering. In practice, multiple scattering can effectively be reduced by decreasing the sample thickness.

Affected by multiple scattering are, in particular, porous materials with high electron density (e.g., graphite, carbon fibers). The multiple scattering of isotropic two-phase materials is treated by LUZATTI [81] based on the Fourier transform theory. PERRET and RULAND [31,82] generalize his theory and describe how to quantify the effect. For the simple structural model of DEBYE and BUECHE [17], RULAND and TOMPA [83] compute the effect of the inevitable multiple scattering on determined structural parameters of the studied material.

In the experiment, multiple scattering is causing a characteristic change of the primary beam profile (Fig. 7.1). As multiple scattering is slowly increasing, shoulders are growing on both sides of the primary beam profile. Nevertheless, in the central part still the original shape of the primary beam profile is conserved (Fig. 7.1b). In the case of strong multiple scattering, the original beam profile is considerably broadened and distorted as a whole (Fig. 7.1c).

---

[5] Such parameters are: primary beam intensity, attenuated primary beam intensity, time, exposure, and the experimental parameters, e.g., sample temperature, force, cross-head position ...

[6] If the experience in programming is high, it is profitable to have the freedom of a programming system like *pv-wave* or IDL.

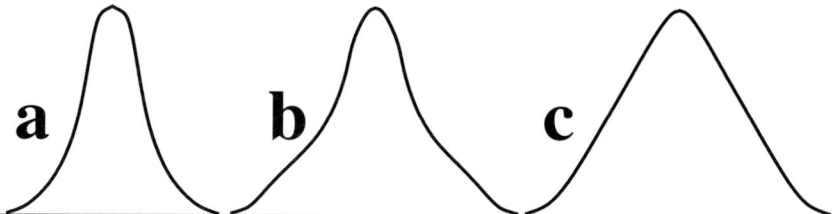

**Figure 7.1.** Variation of a primary beam profile caused by multiple scattering. (a) Unaffected primary beam. (b) Moderate multiple scattering. (c) Strong multiple scattering

In practice, a simple experiment can be performed to test the presence of multiple scattering. First the power of the primary beam is attenuated[7] sufficiently, so that the direct beam can be measured by the detector without the risk of damage. Thereafter, the primary beam is measured once with the sample and once without it. If both images are proportional to each other over the whole range, the effect of multiple scattering is negligible.

## 7.3 Normalization

Divide every scattering pattern and every background by the actual incident flux measured by the first ionization chamber, and divide it by the actual exposure time.

## 7.4 Valid Area Masking

For USAXS and SAXS studies in normal-transmission geometry it is more convenient to carry out this step later – after the absorption and background correction.

Determine the area on the detector, in which valid data have been recorded. If the raw data are a scattering curve, this is normally a single interval. Then remove all invalid points. If the raw data are a scattering pattern, then the beam stop, its holder, the edge of the vacuum tube, and the edge of the detector encircle the valid area. Use the methods of Digital Image Processing (Sect. 2.9) to define the valid area. The result is a *mask image* that can be used for the processing of all data recorded with the same beamline alignment. In this respect processing means multiplication of raw images by the mask. The results are images in which all invalid pixels are pulled down to zero intensity.

---

[7]This is most easily done at a laboratory source where the current of the X-ray tube is decreased to the lowest possible value. At a synchrotron beamline this is more complicated, because the measurement of the primary beam requires special adjustment. So, technically this should be done before the final optical adjustment of the device, as long as the slits can be narrowed for the purpose of intensity attenuation and as long as the primary beam stop is not yet mounted. It is not advised to use absorbers that are mounted behind the monochromator, because they change the spectral composition of the X-ray beam.

## 7.5 Alignment

For USAXS and SAXS studies in normal-transmission *geometry* it is more convenient to carry out this step after the correction for absorption and background, because in these cases absorption and background correction are no function of the scattering angle.

For other geometries the center of the scattering pattern, its orientation, the pixel size of the detector and its distance from the sample must be operated first in order to align the scattering pattern.

If scattering *curves* are processed, the center is simply "a channel number" of the detector and centering is accomplished by subtracting this channel number from all other channel numbers.

For scattering *patterns* the corresponding procedure is more involved and is accomplished by moving the center of gravity of the primary beam into the center of the image matrix. Additionally, if the scattering pattern is anisotropic, its meridian should be aligned in vertical direction. Thus the parameter set of this operation is made from the position of the true center, $(x_c, y_c)$, on the raw image measured in raw pixel coordinates and from an angle of image rotation, $\phi$. If these parameters are known and the sample does not rotate during the experiment, all frames of the experiment can be centered and aligned using the same set.

Interactive rotation and movement utilizing a cross-hair cursor and a system of concentric rings appears to be an easy and reliable way to determine good alignment parameters. Anyway, the cross–hair-and-ring visualization together with the underlying scattering pattern provides a simple method to check the reliability of any automatic or semi-automatic alignment algorithm that can be imagined[8].

If valid areas of images shall be moved and rotated, provisions must be made that during this operation the *valid area does not collide* with the physical edges of the matrix. This can easily be accomplished by sufficiently widening the physical matrix before to move the valid area. Thus the necessary percentage of image widening is a further parameter of an automatic "alignment script".

If the principal *axis* of the sample is *moving* during the experiment, then every frame has to be aligned individually while proceeding through the sequence of frames. In this case it is helpful if the interactive procedure remembers the last set of alignment parameters and the operator only invokes incremental corrections, whenever they become necessary.

## 7.6 Absorption and Background Correction

**The Effect.**   The amount of matter irradiated by the X-ray is varying – as a function of the beamline geometry, the scattering angle, the sample thickness, and sample size. Besides that matter is causing both scattering and absorption of the ray. Scattering intensity and absorption may be a function of the scattering angle for certain

---

[8]I have written and tested several algorithms, but have not yet found a general-purpose and reliable automatic one.

setup geometries. The corresponding relations are readily obtained by simple geometric reasoning (ALEXANDER [7], p. 69-72; GUINIER [6], p. 101-116, p. 180-181). When we follow the route of older textbooks we should consider that the former deductions for reflection geometries are considering free-standing samples, whereas today's focus has turned to thin layers mounted on substrates.

**The Importance.** Absorption and background effects must be properly corrected if the study is aiming at the analysis of *fall-off laws* of diffuse scattering (fractal structure, density fluctuations within phases, transition zones between phases). They should be approximately corrected if absolute changes of the invariant shall be discussed. Absorption and background correction is less important if the study is aiming at topology (i.e., size and arrangement of domains in the sample).

**The Effort of Correction.** If the measurement is carried out in a transmission geometry, an absorption correction can be carried out with fair to high significance if the sample thickness is close to the so-called optimum thickness. If a reflection geometry is chosen, the correction is only simple if the approximation of infinite thickness is allowed. Nevertheless, the so-called linear absorption coefficient must be known – either from direct measurement in transmission, from computation (cf. Sect. 7.6.4), or from a fit in a regression program utilizing a complete model that considers both structure and absorption to fit the scattering data directly.

### 7.6.1 Absorption – the Principle

X-rays are absorbed whenever they travel through matter. As a result, the total transmitted intensity, $I_t$, measured after passing the absorber is only a fraction, $I_t/I_0$, of the incident intensity, $I_0$. For amorphous or polycrystalline material the incremental absorption within a layer of thickness $d\ell$ is constant, and by integration along the complete light path $\ell$ through the sample the absorption law

$$I_t(\ell) = I_0 \exp(-\mu\ell) \tag{7.1}$$

is readily obtained. Here $\mu$ is the linear absorption *coefficient*, which is a function of the X-ray wavelength $\lambda$, the chemical composition of the material, and its mass density (cf. Sect. 7.6.4). $\exp(-\mu\ell)$ is the linear absorption *factor*.

### 7.6.2 Absorption in Normal-Transmission Geometry

Normal transmission geometry is the most frequent setup used in synchrotron beamlines and rotating anode laboratory equipment (cf. Fig. 4.1, Fig. 4.2, and ALEXANDER [7] p. 69-70). Sample, detector, and beam are not moved. The deduction of the absorption relation starts from geometric consideration (Fig. 7.2). The sketch indicates the path of a partial beam of photons that enter the sample through the area $F$, then are scattered after a path $x$ by the angle $2\theta$ so that they have to travel the distance $b$, before finally leaving the sample. It is assumed that each photon is scattered only once, i.e., multiple scattering is excluded. The total length $\ell(t, 2\theta) = x + b$

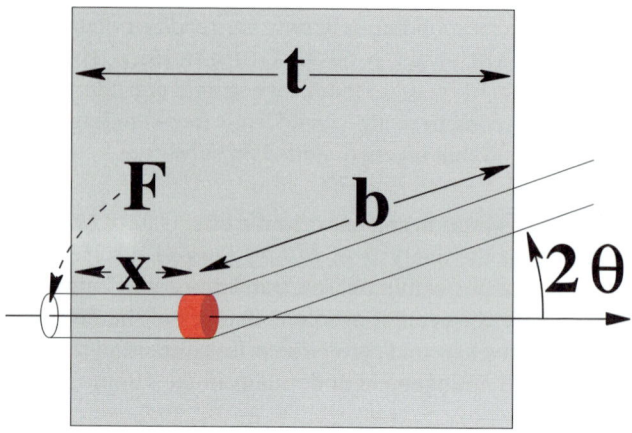

**Figure 7.2.** Absorption in normal-transmission geometry. The path of the photon through a sample of thickness $t$ before and after its scattering about the angle $2\theta$

is established from the sketch. Using Eq. (7.1) and integrating over all $x \in [0, t]$ we obtain for the transmitted intensity the general relation[9]

$$I_t = I_0 F \int_0^t \exp\left(-\mu\ell(x)\right) dx$$

$$= I_0 F \exp\left(-\frac{\mu t}{\cos 2\theta}\right) \frac{1 - \exp\left(-\mu t\left(1 - 1/\cos 2\theta\right)\right)}{\mu\left(1 - 1/\cos 2\theta\right)}. \tag{7.2}$$

With little error the exponential in the numerator of the rightmost fraction of Eq. (7.2) is expanded in a Taylor series resulting in

$$I_t = I_0 F t \exp\left(-\frac{\mu t}{\cos 2\theta}\right). \tag{7.3}$$

We observe that $F t = V$ is the irradiated volume. The maximum of $I_t$ is found at

$$t_{opt} = \frac{\cos 2\theta}{\mu}. \tag{7.4}$$

$t_{opt}$ is the optimum sample thickness in normal-transmission geometry[10]. In SAXS and USAXS we finally have $2\theta \to 0$ and thus $\cos 2\theta \to 1$. Therefore, further simplification

$$I_t = I_0 F t \exp\left(-\mu t\right) \tag{7.5}$$

---

[9]This treatment simply assumes that the material is a homogeneous medium with a scattering power of unity. Thus relative changes are correctly described. In order to put the results back on an absolute scale $I_t (s = 0)$ must be considered.

[10]Equation (7.4) is also valid in symmetrical transmission geometry (ALEXANDER [7] p. 71-72), which is a classical geometry for goniometers equipped with zero-dimensional detectors

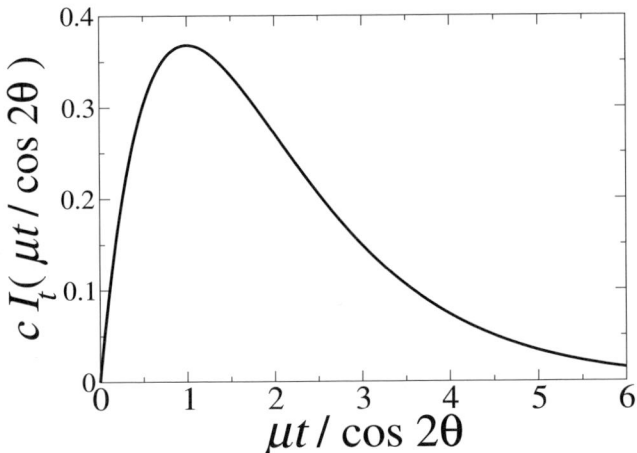

**Figure 7.3.** Effect of absorption in normal-transmission geometry. The total transmitted *scattering* intensity, $I_t$, as a function of the reduced sample thickness. The highest scattering signal is obtained at $\mu t_{opt}/\cos(2\theta) = 1$ with $t_{opt}$ being the optimum sample thickness

is allowed for small scattering angles. The different approximations describe the scattering intensity on the detector as a function of the sample thickness and the scattering angle.

The relation is sketched in Fig. 7.3. From the shape of the curve we anticipate that good scattering signals are obtained, if the thickness of the transmitted sample is in the range $0.5/\mu < t < 3/\mu$.

Moreover, the intensity is additionally increasingly dampened with increasing scattering angle. The corresponding absorption and background correction

$$I_{corr} = I_t' \exp\left(\frac{\mu t}{\cos 2\theta}\right) - I_b$$

is established by Eq. (7.3). Here $I_t'$ is the measured sample scattering pattern on the detector and $I_b$ is the measured pattern of the parasitic background. According to the first-step of pre-evaluation (Sect. 7.3), $I_t'$ and $I_b$ have already been normalized for the incident flux $I_0 F$. After additional division by the sample thickness $t$ we thus have

$$\frac{I_{corr}}{t} = \frac{I}{I_0 V}$$

a scattering pattern that is proportional to the scattering intensity in absolute units, $I/V$. If $\mu$ is known to be constant during the experiment, the *thickness t* or at least a quantity that is proportional to $t$ can be computed from a measured absorption factor $\exp(-\mu t)$.

**The Experimental Determination of the Absorption Factor**  is based on two flux measurements by means of ionization chambers, one placed before ($I_1$), and the other behind ($I_2$) the sample. The second chamber may be replaced by a pin-diode. If $I_{1,0}$ is the reading of the first ionization chamber during a measurement of parasitic background and $I_{1,s}$ is the reading during sample measurement with the analogous nomenclature for the reading of the second ionization chamber then

$$\exp\left(-\mu t\right) \approx \frac{I_{2,s} I_{1,0}}{I_{2,0} I_{1,s}} \tag{7.6}$$

is approximately valid. The measurement of the incident flux $I_1$ is only necessary if the flux is varying like at common synchrotron radiation sources. If the exposure time is low and the sample is thin, the determined absorption factor will not be accurate. In this case the machine background cannot be operated properly. The consequences have been described at the beginning of Sect. 7.6.

### 7.6.3 Absorption in Reflection Geometries

If the X-rays reflected from a large sample are detected and the sample is thicker than $3/\mu$, the assumption of *infinite thickness* is justified. Then the equation for the intensity $I_t$ transmitted into the detector enjoys peculiar simplicity, because it is only a function of the effective irradiated volume

$$V = \frac{F_0}{\mu\left(1 + \sin\alpha_i / \sin\alpha_e\right)}$$

(GUINIER [6], p. 181), with the cross-section, $F_0$, of the incident X-ray beam, the angle of incidence on the sample surface, $\alpha_i$, and the angle of exit with respect to the sample surface being $\alpha_e$. For *symmetrical-reflection geometry* ($\alpha_i = \alpha_e = \theta$) the irradiated volume becomes $F_0/(2\mu)$, and $1/2\mu$ is the penetration depth into the sample. We thus have

$$I_t = \frac{F_0}{\mu\left(1 + \sin\alpha_i / \sin\alpha_e\right)} I_0 \tag{7.7}$$

for the relative variation of the intensity transmitted into the detector. There is no need to subtract a parasitic background, because whatever our detector might have seen when there was no sample – it is completely absorbed in the sample of infinite thickness. In symmetrical reflection

$$I_t = \frac{F_0}{2\mu} I_0 \tag{7.8}$$

the flux $F_0 I_0$ of the primary beam is the only parameter that controls the intensity in the detector.

When reflection geometries are set up in modern scattering applications to study the structure of thin layers, the simplifying assumption of infinite sample thickness is not allowed, and the absorption correction becomes more difficult. Moreover, symmetrical-reflection geometry is utilized less frequently than asymmetrical-reflection geometry with fixed incident angle. Thus both cases are of practical interest.

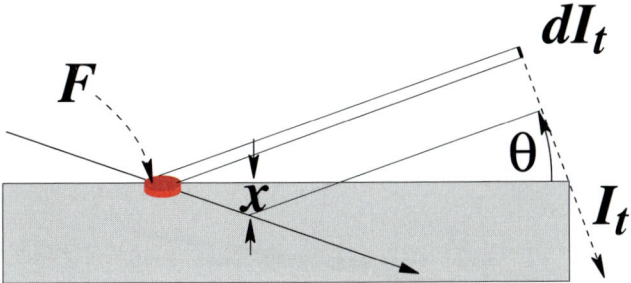

**Figure 7.4.** Sketch for the deduction of the intensity, $I_t$, transmitted into the detector for symmetrical-reflection geometry. The photon is scattered in a depth of $x$. Integration direction is indicated by a straight dashed arrow

### 7.6.3.1 Thin Samples in Symmetrical-Reflection Geometry

Figure 7.4 presents a sketch for the deduction of the intensity in symmetrical-reflection geometry. $F$ is the footprint area of an incident microbeam on the sample surface, and $dI_t$ is the related contribution to intensity. Again utilizing the absorption law Eq. (7.1) we have

$$dI_t = I_0 F(\theta) \exp(-\mu\ell(x)), \tag{7.9}$$

with $\theta$ being both the angle of incidence and the scattering angle, $\mu$ the linear absorption coefficient, and $x$ the depth of interaction between photon and matter. In order to obtain the total intensity $I_t$ transmitted into the detector, we integrate along the dashed straight line and extract from the sketch the path length of the photon $\ell(x) = 2x/\sin\theta$ through the sample[11], introduce the beam cross-section $F_0$ by $F(\theta) = F_0/\sin\theta$, and obtain

$$I_t = I_0 \frac{F_0}{\sin\theta} \int_0^t \exp\left(\frac{-2\mu x}{\sin\theta}\right) dx. \tag{7.10}$$

The integrated result (ALEXANDER [7] p. 81)

$$I_t = I_0 \frac{F_0}{2\mu} A_{sr,1} \tag{7.11}$$

is dominated by

$$A_{sr,1} = \left(1 - \exp\left(\frac{-2\mu t}{\sin\theta}\right)\right), \tag{7.12}$$

the absorption factor for symmetrical reflection in "case 1" according to RULAND & SMARSLY [84], which means that the primary-beam footprint $F$ must not be clipped by the finite sample surface. When, on the other hand, the scattering angle becomes very shallow, a transit of scattering conditions to "case 2" [84] takes place.

---

[11] In close analogy to a deduction of Bragg's law

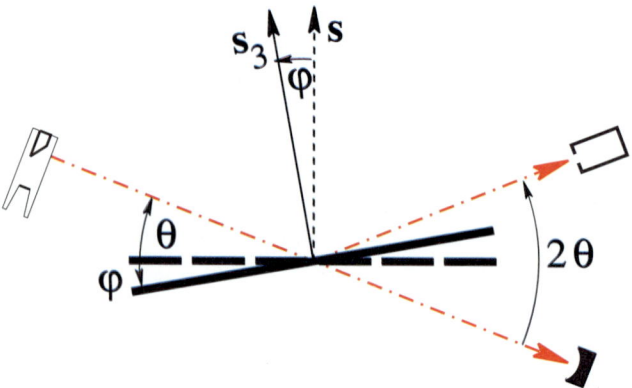

**Figure 7.5.** Relationship between symmetrical ($\varphi = 0$) and asymmetrical ($\varphi \neq 0$) reflection geometry. Bold bars symbolize the sample in symmetrical (dashed) and asymmetrical (solid) geometry. Incident and scattered beam are shown by dashed-dotted arrows. the incident angle is $\alpha = \theta + \varphi$. In any case the intensity is measured at scattering vector **s**. For the tilted sample the sample-fixed scattering vector $\mathbf{s}_3$ is indicated (after [84])

Thus, for thin samples we have to consider a parasitic scattering background originating from the substrate or the sample holder material. The substrate is not illuminated by the full flux, but only by $I_0' = I_0 \exp\left((-\mu t)/(\sin\theta)\right)$, and the resulting background scattering is attenuated once more on its way back through the sample. Thus, if we have measured the background scattering pattern $I_b$ of the pure substrate, and $I_t'$ is the scattering pattern of the thin layer of thickness $t$ on the substrate, then the background corrected intensity $I_t$ is

$$I_t = I_t' - I_b \exp\left(\frac{-2\mu t}{\sin\theta}\right). \tag{7.13}$$

For *very thin sample* thickness $t$ and a scattering angle $2\theta$ that is well above the critical angle of total reflection, the exponential factor is approximately unity and a simple background subtraction without consideration of absorption is allowed. Symmetrical-reflection geometry is only a special case of asymmetrical-reflection geometry.

### 7.6.3.2 Thin Samples in Asymmetrical-Reflection Geometry

Asymmetrical-reflection geometry (GUINIER [6], p. 180-181; RULAND & SMARSLY [84]) has historically been treated in the field of texture analysis. Today it is important for experiments in grazing-incidence geometry. The geometrical relationship is shown in Fig. 7.5. If asymmetrical-reflection geometry is chosen, the incident angle $\alpha = \theta + \varphi$ is frequently small. The geometrical considerations [84] are somewhat more complicated, but nevertheless very similar to the symmetrical geometry. Instead of Eq. (7.11) it follows

$$I_t = I_0 \frac{F_0}{\mu} A_{ar,1} \tag{7.14}$$

with

$$A_{ar,1} = (1 - \cot\theta \tan\varphi) \left[ 1 - \exp\left( \frac{-2\mu t \sin\theta \cos\varphi}{\cos^2\varphi - \cos^2\theta} \right) \right], \tag{7.15}$$

the absorption factor for asymmetrical-reflection geometry. Again, the absorption factor $A_{ar,1}$ is only valid as long as the footprint of the primary beam on the sample is still smaller than the sample surface itself (case 1).

On the other hand, if the primary beam is illuminating the complete sample surface (case 2), the absorption factor for *symmetrical*-reflection geometry becomes

$$A_{sr,2} = \frac{\sin\theta}{\sin\theta_0} A_{sr,1}, \tag{7.16}$$

with $\theta_0$ being that incident angle at which the footprint of the primary beam is just the size of the sample. Therefore the equation is only valid for $\theta < \theta_0$.

For *asymmetrical*-reflection geometry the absorption factor is changed as well, as the primary beam is illuminating the complete sample surface. In this case

$$A_{ar,2} = \frac{\sin(\theta_k + \varphi)}{\sin(\theta_k + \varphi_0)} A_{ar,1} \tag{7.17}$$

is obtained for $\varphi > \varphi_0$, with $2\theta_k$ the constant scattering angle and $\theta_k + \varphi_0$ the angle of incidence at which the footprint of the primary beam equals the surface of the sample. The original paper [84] demonstrates the method how to determine these limiting angles from the geometry of the beamline and the divergence of the beam. If, for different areas of the raw scattering pattern, the illuminating conditions are different, the absorption correction has to consider the transition from case 1 to case 2 by *combining* the respective equations.

### 7.6.4 Calculations: Absorption Factor, Optimum Sample Thickness

The absorption factor of an amorphous or polycrystalline material is computed by summation of incremental contributions from each atom. Thus it is easily computed.

For all chemical elements, mass absorption coefficients $\mu/\rho$ are tabulated [13, 85] as a function of the X-ray wavelength. Chemical composition, mass density $\rho$, and thickness $t$ of the sample are known.

**Example: PET.** Let us consider poly(ethylene terephthalate) (PET, $[C_{10}H_8O_4]_n$, $\rho_{PET}=1.35$ g/cm$^3$) of $t_{PET} = 2$ mm thickness and an X-radiation wavelength $\lambda = 0.15418$ nm (CuK$\alpha$). We set up a table with one row for each chemical element and sum both the masses and the mass absorption coefficients multiplied by the masses. After normalization to the molecular mass of the PET monomer, 192.17 amu, we find $(\mu/\rho)_{PET} = 1291.97/192.17$ cm$^2$/g a value 6.72 cm$^2$/g. Considering the density $\rho_{PET}$ we find for the linear absorption coefficient $\mu_{PET} =$

**Table 7.1.** Scheme for the calculation of the absorption factor

| Atom | $M$ | $m$ [amu] | $M \times m$ | $\frac{\mu}{\rho}$ [cm²/g] | $M \times m \times \frac{\mu}{\rho}$ |
|------|-----|-----------|--------------|----------------------------|--------------------------------------|
| C    | 10  | 12.011    | 120.11       | 4.60                       | 552.51                               |
| H    | 8   | 1.008     | 8.064        | 0.435                      | 3.51                                 |
| O    | 4   | 15.999    | 63.996       | 11.5                       | 735.95                               |
| Sum  |     |           | 192.17       |                            | 1291.97                              |

$(\mu/\rho)_{PET}\, \rho_{PET} = 9.08$ cm$^{-1}$. Thus, for the absorption factor of a sample of 2 mm thickness valid in normal-transmission geometry it follows $\exp(-\mu t) = 0.163$.

The optimum sample thickness for PET of a mass density $\rho_{PET}=1.35$ g/cm$^3$ in *transmission* geometry thus is $t_{opt,PET} = 1/\mu_{PET} \approx 1$ mm. If measured in *reflection*, the PET sample should be at least 3 mm thick.

### 7.6.5 Refraction Correction

If SAXS is measured in reflection (e.g., SRSAXS), one may have to consider the influence of refraction on the observed scattering angle. In particular, when measurement is performed at ultra-small scattering angles (USAXS) the critical angle of total reflection $\theta_c$ cannot be neglected with respect to the scattering angle $\theta$.

For *symmetrical-reflection geometry* the modulus of the true scattering vector is

$$s = \sqrt{s_{obs}^2 - s_c^2},$$

with $s_{obs}$ indicating the modulus of the measured scattering vector. $s_c \approx 2\theta_c/\lambda$ is a very good approximation.

For *asymmetrical-reflection geometry* the relation is more complicated. Considering the geometry sketched in Fig. 7.5, the true tilt angle is

$$\varphi = (\alpha - \beta)/2.$$

For the true scattering angle we have

$$\theta = (\alpha + \beta)/2,$$

in which

$$\alpha = \sqrt{(\theta_k + \varphi_{obs})^2 - \theta_c^2}$$

and

$$\beta = \sqrt{(\theta_k - \varphi_{obs})^2 - \theta_c^2}$$

are expressed by the tuned scattering angle, $\theta_k$, the measured tilt angle, $\varphi_{obs}$, of the normal to the sample, and by the critical angle of total reflection, $\theta_c$.

## 7.7 Reconstruction of Proper Constitution

We are processing noisy data. Thus, intensities may have become negative by accident. In order to mark such spots as invalid data, they should be set to zero. This is accomplished by the masking formalism of image processing

```
wave> m=img.map GT 0
wave> img.map=img.map*m
```

The first line generates the mask of the still valid pixels, the second pulls the invalid pixels to zero level. The example works in *pv-wave* and IDL, if img.map is the map of the image.

## 7.8 Conversion to Reciprocal Space Units

### 7.8.1 Isotropic Scattering

For isotropic scattering patterns, the relation between the channels or pixels on the detector is simply given by Eq. (2.5), and an *s*-value is readily associated to each pixel of the detector. If a 2D detector has been used to record the data, the signal-to-noise ratio can thus be significantly enhanced by averaging all pixels with the same *s*-value[12] (cf. Sect. 8.4.1).

### 7.8.2 Anisotropic Scattering

#### 7.8.2.1 USAXS and SAXS

For USAXS and SAXS data the tangent-plane approximation is valid and the relation between scattering angle and the units of reciprocal space are given by Eq. (2.7). If the scattering pattern is properly aligned with the vertical direction identical to a fiber axis or the polymer chain direction, then $s_y = s_3$. In similar manner the $s_x$-axis of the detector is related to the actual orientation of the sample with respect to the beam.

#### 7.8.2.2 MAXS and WAXS with Fiber Symmetry

For MAXS and WAXS the problem is more involved. If the scattering pattern shows fiber symmetry, the considerations of Sect. 2.8.2.2 apply.

#### 7.8.2.3 MAXS and WAXS Without Fiber Symmetry

This is the general case of texture analysis, in which the sample must be tilted and rotated in order to collect all the data required for a complete quantitative data analysis [49].

---

[12]These pixels are found on concentric rings about the center of the pattern.

## 7.9  Harmony

As long as there is at least uniaxial symmetry and the fiber axis is in the detector plane, the scattering pattern can be split into four quadrants which should carry each identical information. This means that there is some harmony in the scattering pattern, from which missing data can be reconstructed[13].

After this reconstruction the scattering pattern should be smooth. If, on the other hand, "seams" are observed at the edges of the former invalid regions this shows that penumbra was not detected and the mask of the valid pixels was chosen too large. Solution: erode the old mask and return to the start of pre-evaluation.

## 7.10  Calibration to Absolute Scattering Intensity

Calibration to absolute intensity means that the scattered intensity is normalized with respect to both the photon flux in the primary beam and the irradiated volume $V$. Thereafter the scattering intensity is either expressed in terms of electron density or in terms of a scattering length density. Both definitions are related to each other by COMPTON's classical electron radius.

**Fields of Application.**  In SAXS a calibration to absolute intensity is required if extrapolated or integrated numerical values must be compared on an absolute scale. Examples are the determination of density fluctuations or the density difference between matrix and domains as a function of materials composition.

In WAXS of soft condensed matter, studies of the intensity in absolute units are not common, unless the method for the exact determination of X-ray crystallinity according to RULAND is applied (cf. Sect. 8.2.4).

**General Routes.**  If a SAXS beamline in normal transmission geometry is used, calibration to absolute intensity is, in general, carried out *indirectly* using secondary standards. *Direct* methods require direct measurement of the primary beam intensity under consideration of the geometrical setup of the beamline. On a routine basis such direct calibration was commercially available for the "historic" Kratky camera equipped with zero-dimensional detector and "moving slit device"[14].

### 7.10.1  The Units of Absolute Scattering Intensity

The radiation strength of a single electron is determined by elementary quantities that need not be considered in materials science. These quantities and the characteristics of the detector are automatically considered if the primary beam intensity is directly measured or indirectly determined by means of a secondary standard.

---

[13]My *pv-wave* procedures sf_harmony.pro and sf_fillharmony.pro
[14]By means of such equipment secondary standards can be made.

**Electron Density.**   Continuing the preceding considerations, calibration to absolute intensity means normalization to "the scattering of a single electron", $I_e$ that can be expressed in electron units, [e.u.]. Inevitably a calibration to absolute units involves also a normalization with respect to the irradiated volume $V$. Thus, for the field of materials science a suitable dimension of the absolute intensity is $[I/V] =$ e.u./nm$^3$ – "The intensity measured in the detector is originating from a material with an average electron density of 400 electrons per nanometers cubed". The electron density itself is easily computed from mass density and chemical composition of the material (cf. Sect. 2.2.1).

**Scattering Length Density.**   In a more fundamental definition, absolute intensity is expressed in terms of a scattering length density. It is suitable in particular, if data from X-ray scattering experiments shall be compared to data from neutron scattering – a field, in which scattering length density is the natural unit of absolute intensity. In X-ray scattering the COMPTON *classical electron radius*

$$r_e = \frac{e^2}{m_e c^2} = 2.818 \times 10^{-15} \, \text{m} \qquad (7.18)$$

is identified as the THOMSON *scattering length* of a single electron[15]. Finally the electron is replaced by its scattering cross-section (e.u.$\rightarrow \sigma_e = r_e^2$), and the dimension of the absolute intensity becomes $[I/V] =$ nm$^{-1}$.

### 7.10.2 Absolute Intensity in SAXS

For the calibration to absolute intensity several direct and indirect methods have been proposed (FEIGIN and SVERGUN [86], p. 73-76).

#### 7.10.2.1 The Idea of Direct Calibration

In normal transmission geometry[16] any mathematical treatment of calibration to absolute units [87–90] starts from the basic *differential* relation among the scattering intensity in the detector, the primary intensity and the structure

$$dP = i_0 f_p \frac{r_e^2}{r^2} d\sigma_s t \exp(-\mu t) \, i_{dV}(\mathbf{s}) \, d\sigma_c, \qquad (7.19)$$

with $dP$ as the differential power scattered out of an irradiated volume element $d\sigma_s t = dV$ of the sample into an area element $d\sigma_c$ of the detector. $i_0$ is the intensity of the primary beam, $f_p$ the polarization factor, $r_e$ the classical electron radius (Eq. (7.18)), $r$ the distance between $d\sigma_s$ and $d\sigma_c$. $dV$ is the volume element of the sample that is irradiated by the considered differential primary beam element, and $i_{dV}(\mathbf{s})$ is the Fourier transform of the correlation function $\rho^{*2}(\mathbf{r})$ restricted to $dV$

---

[15] Here $e$ and $m_e$ are charge and mass of the electron, respectively. $c$ is the velocity of light

[16] For other geometries the term $t \exp(-\mu t)$ must be redefined as demonstrated with the treatment of absorption in Sect. 7.6.

$$i_{cV}(\mathbf{s}) = \frac{i(\mathbf{s})}{dV} = \frac{1}{dV} \int_{dV} \rho^{*2}(\mathbf{r}) \exp(2\pi i \mathbf{r} \mathbf{s}) d^3 r. \qquad (7.20)$$

This is the *differential* definition of the absolute intensity. The *total* absolute intensity can be deduced by integration from Eq. (7.19) and Eq. (7.20) for any normal transmission *geometry*. Geometries are discriminated by the shape and size of the irradiated volume, the image of the primary beam in the registration plane[17] of the detector, and the dimensions of the detector elements[18].

### 7.10.2.2 Direct Calibration for the Kratky Camera

For the classical slit-focus camera the result of the integration is [89]

$$\frac{J(\mathbf{s})}{V} = \frac{R}{r_e^2 \lambda t \exp(-\mu t) H L} \frac{P_R(\mathbf{s})}{P_0'}. \qquad (7.21)$$

$J(\mathbf{s}) = \int I(\mathbf{s}) ds_2$ is the slit-smeared scattering intensity, $P_0'$ is the total primary beam intensity per slit-length element – a quantity determined by the moving slit device. $R$ is the distance between sample and detector slit as measured on the optical axis of the camera. $L$ is the (fixed and known) length of the detector slit in the registration plane. $H$ is the (adjustable) height of the detector slit. $\exp(-\mu t)$ is the linear absorption factor of the sample[19].

In Eq. (7.21) the normalization to the scattering cross-section $r_e^2$ leads to the definition of absolute intensity in electron units which is common in materials science. If omitted [90, 91], the fundamental definition based on scattering length density is obtained (cf. Sect. 7.10.1).

**A Practical Hint**[20].    In order to most accurately determine $H$ in Eq. (7.21), a mathematical theorem concerning convolution of a function with a shape function are helpful. The measured primary beam profile of the Kratky camera

$$h_B(x) = h_{B0}(x) * Y_H(x) \qquad (7.22)$$

is the convolution of the intrinsic beam profile "before the detector" $h_{B0}(x)$ with the slit height profile $Y_H(x)$ – a shape function. Here $x$ is the length coordinate[21] in the plane of registration. With $H$ the sought-after integral breadth of the slit and $B_{B0}$ the integral breadth of the intrinsic beam profile, it follows from Eq. (7.22) for the integral breadth of the observed primary beam profile

---

[17] The registration plane of the detector is the plane in which the pixels of the detector image are shaped
[18] For a 2D detector this is height and width of the detector pixels. For the Kratky camera with zero-dimensional counter this is the height and length of the measuring slit.
[19] The division by $t \exp(-\mu t)$ eliminates effects of absorption and sample thickness. The polarization factor is 1 in SAXS.
[20] This procedure can only be applied for a Kratky camera with zero-dimensional detector. It shows the value of this classical step-scan device for studies of scattering in absolute intensity units.
[21] measured in micrometers

$$B_B = \begin{cases} H & / H > B_{B0} \\ B_{B0} & / H < B_{B0} \end{cases}. \qquad (7.23)$$

It is thus reasonable to make the slit height $H$ of the detector slit wider than the integral breadth of the intrinsic primary beam profile. In this case the observed integral breadth equals $H$ – and can be accurately determined from the measured primary beam profile.

**Example: A Calibration for Kratky Camera Data.** Let us assume that the measured integral breadth of the primary beam profile is $H = 223 \times 10^3$ nm. Due to proper alignment this is the height of the detector slit. Its length is $L = 1.6 \times 10^7$ nm. The distance between sample and detector is $R = 2 \times 10^8$ nm. Cu$k_\alpha$–radiation with $\lambda = 0.15418$ nm is used. $P_0'$ is determined from a measurement with the moving slit device according to

$$P_0' = \frac{N v}{l_1 l_2 i}.$$

Here $N = 140050$ is the total number of registered pulses during the measurement with empty sample holder, $v = 0.208$ cm/s is the velocity of the moving slit during the scan of the primary beam, $i = 16$ is the number of scans[22], $l_1 = 32 \mu$m is the slit opening of the moving vertical slit, and $l_2 = 100 \mu$m is the slit opening of a fixed vertical slit mounted in the plane of registration. The values of the openings must be measured under a microscope. Thus,

$$P_0' = 2.845 \, s^{-1} nm^{-1}$$

is obtained. Combined with the geometrical constants we have

$$\frac{H L}{r R} P_0' = 50.76 \, 10^3 \, s^{-1}.$$

The absorption factor of the studied sample is measured by means of the moving slit device, as well. For this purpose the sample is mounted in the sample holder and the moving slit measurement is performed. We measure $N_S = 50031$ in 16 scans. Because the heights of the moving and the fixed vertical slits are 2 cm, the moving slit registers the SAXS as well – as is required from the definition of absorption. Then

$$\exp(-\mu t) = \frac{N_s}{N} = \frac{50031}{140050} = 0.3572$$

is computed. If the sample thickness $t$ cannot be determined with sufficient accuracy, we compute $\mu$ from the chemical composition of the sample (cf. Sect. 7.6.4) and resolve

$$t = -\ln(\exp(-\mu t))/\mu.$$

Thereafter the slit-smeared scattering intensity is readily expressed in absolute units $[J(s)/V] = $ e.u./nm$^4$.

---

[22] chosen high enough to obtain good statistics

### 7.10.2.3 Direct Calibration for a Synchrotron Beamline

Direct calibration to absolute intensity is not a usual procedure at synchrotron beamlines. Nevertheless, the technical possibilities for realization are improving. Therefore the basic result for the *total* scattering intensity measured in normal transmission
geometry is presented. At a synchrotron beamline point-focus can be realized in good
approximation and the intensity $I(\mathbf{s})$ is measured. Then integration of Eq. (7.19) results in

$$\frac{I(\mathbf{s})}{V} = \frac{R}{r_e^2 \lambda t \exp(-\mu t) H L} \frac{P_R(\mathbf{s})}{P_0}, \qquad (7.24)$$

with $H$ and $L$ now denominating height and width of each detector pixel, whereas
$P_0 = N/(t S_0)$ is the power measured in the detector; as measured in $N$ pulses per
time $t$, now normalized not with respect to slit length, but instead to the *integral area*

$$S_0 = \frac{1}{\max(h_0(x,y))} \iint h_0(x,y)\,dxdy$$

of the image $h_0(x,y)$ of the primary beam on the 2D detector[23]. Equation (7.24) is
valid for small and medium scattering angles, as long as the polarization factor is 1.
Again, the division by $r_e^2$ turns the result into units of electron density. The sample is
assumed to be flat and completely covering the cross-section of the primary beam.

**Cylindrical Filaments.**   If cylindrical filaments are studied, the last-mentioned
assumption is not be fulfilled. In this case it is more suitable to cross an elongated[24]
primary beam and the fiber in such a way that the irradiated height of the beam,
$B_{B0}$, is constant. Then the irradiated volume becomes $V = B_{B0}\,\pi r_f^2$, with $r_f$ being the
radius of the fiber. In many practical applications the studied fiber is very thin, and
the effective absorption coefficient can be approximated by $\exp(-\mu t_{eff}) \approx 1$. If this
approximation is not feasible, the absorption factor $\exp(-\mu\ell)$ must be integrated for
the chord length distribution $g_c(\ell)$ of a circle with radius $r_f$ (cf. p 168) in order to
yield the effective absorption.

**Protection of the Detector.**   With all direct calibration methods the primary
beam intensity must be measured. If the primary beam itself is attenuated, shape of
the beam and spectral composition of the radiation may be altered. This problem is
avoided if the load of the detector is reduced by scanning the beam using a slit or a
perforated disc. On the other hand, in order to be useful at a powerful synchrotron
beamline these devices should have very tiny and well-defined slits or holes.

---

[23] $x$ and $y$ are lengths measured "in centimeters" on the detector, **not** measured in units of reciprocal space.
[24] Open the horizontal slits a bit more than usual, if the beam is not wide enough anyway.

### 7.10.2.4 Indirect Calibration Using a Polymer Sample

The so-called "Lupolen® standard"[25] is a well-known secondary standard in the field of SAXS. In conjunction with the Kratky camera it is easily used, because its slit-smeared intensity $J(s)/V$ is constant over a fairly wide range, and this level is chosen as the calibration constant. In point-focus setups the SAXS of the Lupolen standard neither shows a constant intensity region, nor is the reported calibration constant of any use.

A proper calibration constant for any beamline geometry is the invariant $Q$. Thus, the Lupolen standard or any other semicrystalline polymer that previously has been calibrated in the Kratky camera can be made a secondary standard for a point-focus setup, after its invariant $Q$ has been computed in absolute units – based on a measurement of its SAXS in the Kratky camera.

The calibration process then involves measurement of the complete scattering curve of the secondary standard and the evaluation[26] of $k$ by determination of POROD's law with its asymptote $A_P$ and the density fluctuation background $I_{Fl}$, numerical extrapolation of the function $s^2(I(s) - I_{Fl})$ towards $s = 0$, and finally computation of the scattering power

$$k = \int_0^{s_{max}} s^2 (I(s) - I_{Fl}) \, ds + \frac{A_P}{s_{max}} \qquad (7.25)$$

by integration (cf. Sect. 8.4.3). Identification with the known value of $Q$ yields the calibration factor which is required to transform $I(s) \rightarrow I(s)/V$, the scattering intensity to absolute units. Instead of numerical extrapolation to large scattering angle we follow RULAND [92] and use the *analytical continuation* of the integrand given by the Porod law, $c/s^2$. Thus, the remainder term is readily computed

$$\int_{s_{max}}^{\infty} c/s^2 \, ds = c/s_{max}. \qquad (7.26)$$

The measured SAXS curve of the calibration sample must have been preprocessed in the usual way (cf. Sects. 7.3 - 7.6). Therefore it is important to have calibration samples with a well-defined thickness[27]. Because synchrotron beamlines can be adjusted to a fairly wide range of radiation power, it is important to have thin calibration samples for a high-power adjustment (e.g., common SAXS with wide slit openings) and thick calibration samples for low-power adjustments (e.g., USAXS with microbeam). For calibration samples from synthetic polymers, thicknesses ranging between 0.2 mm and 3 mm are reasonable. It appears worth to be noted that not only polymers, but as well glassy carbon [88] can be used as a solid secondary standard for the calibration to absolute intensity.

---

[25]The standard is made from Lupolen® 1800, a branched low-density polyethylene with a very broad long period distribution. Therefore the slit-smeared SAXS peak is only a shoulder that starts with a plateau.

[26]This evaluation is performed in minutes using my computer program TOPAS.

[27]Such samples can be machined by means of a low-speed diamond saw (e.g., Buehler Isomet®).

### 7.10.2.5 Indirect Calibration by Fluid Standards

Pure liquids can be used for the purpose of calibration to absolute intensity, because their diffuse scattering $I_{Fl}(0) = \lim_{s \to 0} I_{Fl}(s)$ caused from density fluctuations can be computed theoretically. Some examples are in the literature [91, 93–95].

Another fluid standard used in the literature is a suspension of colloidal noble-metal particles in a solvent [96]. The method is explained starting on p. 134. The application of such calibration methods is in particular feasible, if polymer solutions are studied and thus the measurement of a calibration fluid does not require to modify the setup.

> **Warning.** From time to time it is postulated in the literature that the intensity value $I_{Fl}(0)$ that is used for the purpose of calibration could, as well, be determined in a slit-focus setup (after Rigaku-Denki or Kratky) either by extrapolation to zero scattering angle [93] or even ("since the slit-smearing of a constant obviously is a constant") directly from the almost constant background [91]. Thereafter this value is related to the theoretical density fluctuation scattering of the calibration fluid. This oversimplification leads to a systematic error which cannot be tolerated with respect to a calibration method [95]. In fact, the fluctuation background $I_{Fl}(s)$ is not constant, but slowly varying as a function of the scattering vector, as is shown both from experiment [91, 93–95] and theory (Sect. 8.3.1). Thus, *computation* of the slit-smearing effect is a necessity. It is, in general, sufficient to model the fluctuation background by
>
> $$I_{Fl}(s) = I_{Fl}(0) + b s^2. \tag{7.27}$$
>
> In this case the result of slit-smearing is [95]
>
> $$J_{e.u.}(s) = I_{e.u.}(0) + b_{e.u.} \sigma_w^2 + b_{e.u.} s^2,$$
>
> with $\sigma_w^2 = \int s_1^2 W(s_1)\, ds_1$ being the 2nd moment of the normalized primary beam profile $W(s_1)$ of the slit-camera in the direction $s_1$ of the slit length.
>
> Obviously the slit-smearing causes an *additional* background $b_{e.u.} \sigma_w^2$, which is a function both of the sample material and the temperature of the sample. Disregarding this background results in a systematic error. For benzene – one of the best-suited calibration fluids – the error is 4% at room temperature. For polymers errors of up to 65% (polystyrene at room temperature) have been verified both theoretically and experimentally.

### 7.10.3 A Link to Absolute Intensity in WAXS

WAXS calibration for polymer materials can be simplified if one considers the fact that polymers are chain molecules. This means that for wide scattering angles in

WAXS the scattering is completely described by the interactions of neighboring atoms *along a single chain*, the so-called *single-chain structure factor*. Cf. descriptions of the RULAND method [14] in textbooks [7, 22].

The study of absolute or integrated intensity is not very common in the field of WAXS of soft matter. Moreover, in this field both the methods (single crystal, mosaic crystal; both with and without averaging by rotating the crystal[28]) and the geometries (of diffractometers, samples and sample structures) vary much more than in the field of SAXS. Because any methodical or geometrical change causes a variation of the equation for the determination of absolute intensity, there is no general final equation for the computation of absolute WAXS intensity. Several cases have been treated by WARREN ([97], Chap. 4). In particular, his treatment of absolute intensity for thick, isotropic powder samples studied in symmetrical reflection ([97], Eq. (4.12)) is valuable for application in soft-matter materials science. Unfortunately it is beyond the scope of this book[29].

On the other hand, modern macromolecular X-ray diffraction of crystals from biological materials (protein crystallography) is frequently carried out in a rotating-crystal setup [98,99] (ALEXANDER [7], p. 51), (WARREN [97], Chap. 7), (GLOCKER [100], p. 250) using plane 2D detectors in normal transmission setup. In this case the absolute scattering intensity of reflection *hkl* is given by the DARWIN equation[30] (cf. WARREN [97], Eq. (4.7))

$$\frac{I_{hkl}}{V_{cr}I_0} = r_e^2 \frac{\lambda^3}{V_u^2} \frac{f_{L\omega}}{\omega} f_P f_A |F_{hkl}|^2, \qquad (7.28)$$

if the volume of the crystal, $V_{cr}$, is completely bathed in a primary beam of intensity $I_0$. The crystal made from unit cells of volume $V_u$ is rotated with an angular velocity $\omega$. $F_{hkl}$ is the structure factor of the reflection. It is, in general, subjected to thermal disorder according to DEBYE-WALLER (cf. p. 109 and WARREN [97], Eq. (3.24)). $f_A$ is the absorption factor, which is a function of the geometry of the sample and the setup (cf. Sect. 7.6). $f_P$ is the polarization factor, which is a function of the radiation source (cf. Sect. 2.2.2). The LORENTZ factor $f_{L\omega}$ takes care of the fact, that by rotating the crystal the intensity of the point-shaped reflection is distributed on a ring in reciprocal space[31]. As presented, Eq. (7.28) returns the absolute intensity in units

---

[28] Spinning a crystal during measurement of WAXS patterns is an old method that turns any scattering pattern into a fiber pattern. The rotational axis becomes the principal axis. Thereafter isotropization of the scattering data is simplified because the mathematical treatment can resort to fiber symmetry of the measured data. In the literature the method is addressed as the *rotating-crystal method* or *oscillating-crystal method*.

[29] The treatment is rather involved, so the reader is asked to consider the WARREN's textbook. In addition, if the concept shall be applied to synchrotron beamline setups with flat 2D detectors some geometrical modifications must be carried out.

[30] Cited by GLOCKER [100]: C. G. DARWIN (1914) Phil. Mag., 27, 315-333

[31] In other fields of scattering the Lorentz factor is defined independently from a forced rotation as $f_L = f_{L\omega}/\omega$. In this more general definition, if the crystal is rotated about the $(00\ell)$- or $s_3$-direction, the radius of the aforementioned ring becomes $s'_{12}$, and the Lorentz factor correspondingly is $f_L = 1/(2\pi s'_{12})$, the inverse of the circumference of the ring. $s'_{12}$ itself is computed from the position of the considered reflection $s_{hkl} = (s'_1, s'_2, s'_3)$ according to $(s'_{12})^2 = (s'_1)^2 + (s'_2)^2$. In the general definition $f_L$ is the inverse

of scattering length density. Electron-density units are obtained after the factor $r_e^2$ has been canceled.

It should be clear that the DARWIN equation with its special LORENTZ-polarization factor as reported by WARREN ([97], Eq. (4.7)) is only valid for unpolarized laboratory sources and the rotation-crystal method. An application to different setup geometries, for example to synchrotron GIWAXS data of polymer thin films is not appropriate.

---

of the area over which the reflection is smeared on the sphere with radius $|s_{hkl}|$. Such smearing may be caused either from forced rotation of a single crystal or from an orientation distribution of crystallites in a mosaic crystal. In the notion of the rotation-crystal method, the plane 2D detector probes the Ewald sphere in reciprocal space, and $f_L$ is the "length of the light pulse on the detector", i.e., the time the peak takes to pass through the Ewald sphere normalized to the cycle time of the forced rotation.

# 8 Interpretation of Scattering Patterns

## 8.1 Shape of the Scattering Intensity at Very Small Angles

### 8.1.1 GUINIER's approximation

GUINIER's law states that for any scattering pattern, $I(\mathbf{s})$, of a diluted system the initial intensity decay is approximated by a Gaussian

$$I(s_i) = \lceil I(\mathbf{s})\rceil_1 (s_i) = I(0) \exp\left(-4\pi^2 R_{gi}^2 s_i^2\right) \tag{8.1}$$

for any direction $s_i$ chosen – as long as $R_{g,i}s_i$ is small enough. For samples that show isotropic scattering the approximation is commonly written

$$I(s) = I(0) \exp\left(-\frac{4\pi^2}{3}R_g^2 s^2\right) \tag{8.2}$$

with $R_g$ called the *radius of gyration* or the *Guinier radius*. For isotropic scattering both Eq. (8.1) and Eq. (8.2) are applicable (cf. Eq. (8.8) and Eq. (8.9)).

Figure 8.1 sketches the analysis according to GUINIER's law. Not applicable is the Guinier approximation, if valid data deviate at low angles (in the sketch: triangles and crosses). Triangles show a deviation, for which an analysis according to

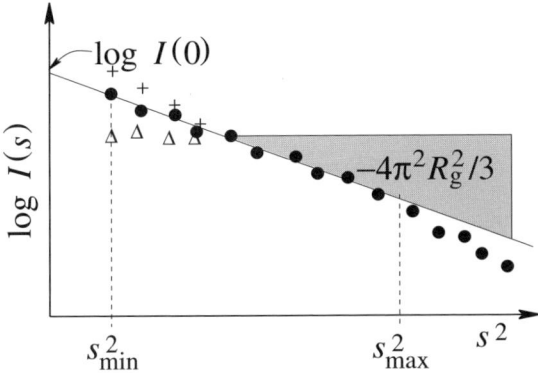

**Figure 8.1.** Guinier plot. Applicability of GUINIER's approximation to scattering data and determination of its parameters, $I(0)$ and $R_g^2$. $s_{min}$ is the lowest scattering angle at which valid data are present. From $s_{max}$ deviation between the data and the straight line is observed

GUINIER's law should never be considered. Crosses show a deviation that may be explained by the influence of the direct primary beam to the measured data. The value of $s^2_{max}$ gives some hint on the confidence of the approximation. After determination of $R^2_g$ the relation $0.1 \leq R^2_g s^2_{max} \leq 0.4$ should hold.

### 8.1.2 Usability for Data Extrapolation

Without any interpretation GUINIER's law can be used to extrapolate small-angle scattering data towards zero scattering angle, if the measured data cover a part of the Guinier region, i.e., the region where Eq. (8.1) or Eq. (8.2) is valid.

### 8.1.3 Usability for Structure Parameter Determination

GUINIER's law exhibits two parameters, $I(0)$ and $R^2_g$, which describe structural aspects of the sample. The experimentalist should consider their determination, if the recorded SAXS data show a monotonous decay that is indicative for the scattering from uncorrelated[1] particles. Particularly useful is the evaluation of GUINIER's law, if almost identical particles like proteins or latices are studied in dilute solution (cf. PILZ in [101], Chap. 8). The absolute value of $I(0)$ is only accessible, if the scattering intensity is calibrated in absolute units (Sect. 7.10.2).

**Guinier Radius.**   The parameter $R_g$ is called *radius of gyration* or *Guinier radius*. The Guinier radius is a measure of particle size. For an ensemble of identical particles, $R^2_g$ is reduced to the second central moment of the orientation-averaged particle density (cf. Eq. 8.4). For the discussion of polydispersity cf. p. 19. The common problems with the evaluation of GUINIER's law are clearly discussed by BALTÁ and VONK ([22], Sect. 7.4), who also present solutions. The bottom line is: materials containing heterogeneous particles exhibit a shortened Guinier region, and the meaning of the parameter $R^2_{gi}$ is changed [21]. If the particle concentration becomes too high, first the prefactor in the exponential function is changed and, finally, even the decay itself becomes distorted. For extremely anisotropic particles (rods, discs) the plain Guinier law is no good approximation (cf. POROD in [101], p. 32-37).

### 8.1.4 Determination of the Parameters of GUINIER's law

The structure parameters $I(0)$ and $R^2_{gi}$ are determined in a linearizing plot of $\log(I(s_i))$ *vs.* $s^2_i$ – the Guinier plot (cf. Fig. 8.1). For isotropic scattering $R^2_{gi} = R^2_g/3$ is identical for every direction $i$. A cross-check after determination can help to indicate problems: if $s_{max}$ is the maximum $s$-value up to which linearity was found in the Guinier plot, then $s_{max}R_g > 0.1$ should be valid (GUINIER in [102], Sect. 5.3) even for the unfortunate case of long rodlike particles, as long as their size does not vary and the effect of correlations among them may be dropped.

---

[1] Strictly speaking, additionally the particles must be convex (i.e., without indentations or holes) and homogeneous (i.e., without density oscillations)

### 8.1.5 Meaning of the Parameters of GUINIER's Law

Assuming that no correlation among particles is present, the meanings of the structure related parameters of GUINIER's law are readily established by application of the mathematical tools of scattering (cf. Chap. 2). Further assumptions state that each particle $k$ is considered to be immersed in a matrix of constant electron density and that its correlation function is monotonously[2] decaying. Thus, the particle is discriminated from the matrix by its number $m_k$ of electrons that it has more (respectively, less) than the homogeneous matrix. $m_k$ is called the *number of excess electrons* of the particle $k$.

**Intensity at s=0.**   The requested intensity $I(0)$ is obtained by integration of the correlation function over the irradiated volume $V$. As the result of this operation, each particle $k$ contributes with its $m_k^2$ to $I(0)$. Thus, with $N$ particles present in the irradiated volume we finally have

$$I(0) = \lim_{s \to 0} I(s) = N \langle m^2 \rangle_V, \tag{8.3}$$

with $\langle m^2 \rangle_V$ the average square of the particle excess electrons[3] in $V$. Whenever scattering intensity is normalized to absolute units, the determined quantity is not $I(s)$ but[4] $I(s)/V$. Correspondingly $(N/V)\langle m^2 \rangle_V$ is computed. $N/V$ is the *number particle density*.

**Anisotropic Particle Scattering: Varying Intensity Decay in Different Directions.**   In case of anisotropy the decay of the scattering intensity $I(\mathbf{s})$ is a function of the direction chosen. The intensity extending from $s = 0$ outward in a deliberately chosen direction $i$ is mathematically the definition of a slice (cf. Sect. 2.7.1, p. 22). Thus, the Fourier–Slice theorem, Eq. (2.38), turns the particle density function $\Delta\rho(\mathbf{r})$ into a projection $\{\Delta\rho(\mathbf{r})\}_1(r_i)$ and the scattering intensity is related to structure by

$$\lceil I(\mathbf{s}) \rceil_1 (s_i) = \mathscr{F}(\{\Delta\rho(\mathbf{r})\}_1(r_i)) \, \mathscr{F}^*(\{\Delta\rho(\mathbf{r})\}_1(r_i))$$

with $\mathscr{F}^*(h(r))$ denoting the complex conjugate of $\mathscr{F}(h(r))$. According to its definition the Fourier transform itself is

$$\mathscr{F}(\{\Delta\rho(\mathbf{r})\}_1(r_i)) = \int_{-\infty}^{\infty} \{\Delta\rho(\mathbf{r})\}_1(r_i) \exp(2\pi i r_i s_i) \, dr_i.$$

After series expansion of the harmonic kernel, term-by-term integration and recombination with its complex conjugate a series expansion of Eq. (8.1) is obtained in which $R_{gi}^2$ is identified as

---

[2]Monotony is not guaranteed for inhomogeneous or non-convex particles, but this poses no principal problem.

[3]The expected relation to the weight-average (excess) molecular weight is established by Eq. (1.4)

[4]Reason: The standard is measured with the same primary beam cross-section as is the studied sample. After division by the intensity of the standard and the thickness of the studied sample, a division by the irradiated volume has been accomplished.

$$R_{gi}^2 = \frac{\int_{-\infty}^{\infty} (r_i - r_{0i})^2 \{\Delta\rho(\mathbf{r})\}_1 (r_i) dr_i}{\int_{-\infty}^{\infty} \{\Delta\rho(\mathbf{r})\}_1 (r_i) dr_i}. \tag{8.4}$$

With $r_{0i}$ the particle's center of gravity, this equation defines $R_{gi}^2$ by the second central moment of the density distribution of the particle projected on a line extending in $r_i$ direction. The equation is simplified ($r_{0i} = 0$) if the origin of the coordinate system is chosen to rest in the center of gravity.

For a system of uncorrelated but highly oriented particles in a sample (e.g., oriented needle-shaped voids in a fiber) it may be possible to factorize the particle density, e.g.,

$$\Delta\rho(\mathbf{r}) = \Delta\rho_1(r_1) \Delta\rho_2(r_2) \Delta\rho_3(r_3).$$

In this case we obtain three principal projected central moments, $R_{g,1}^2$, $R_{g,2}^2$, and $R_{g,3}^2$ – and the anisotropic scattering intensity in the vicinity of zero scattering angle is modeled

$$I(\mathbf{s}) = I(0) \exp\left(-4\pi^2 R_{g,1}^2 s_1^2\right) \exp\left(-4\pi^2 R_{g,2}^2 s_2^2\right) \exp\left(-4\pi^2 R_{g,3}^2 s_3^2\right)$$

by a product of coordinate functions, as well. For small and very small arguments two successive approximations are obvious and of practical interest: first the exponentials are expanded, second the products are eliminated according to $(1 - \varepsilon_1)(1 - \varepsilon_2) \approx 1 - \varepsilon_1 - \varepsilon_2$. The result for the structure with oriented particles is

$$I(\mathbf{s}) = I(0) \left(1 - 4\pi^2 \left(R_{g,1}^2 s_1^2 + R_{g,2}^2 s_2^2 + R_{g,3}^2 s_3^2\right)\right), \tag{8.5}$$

and if the same material has lost its orientation completely, the isotropic intensity decay follows upon solid-angle averaging $\langle\rangle_\omega$ of the intensity

$$\begin{aligned} I(s) &= \langle I(\mathbf{s}) \rangle_\omega \\ &= \frac{1}{4\pi} \int_0^{2\pi} \int_0^{\pi} I(s, \psi, \varphi) \sin\psi \, d\psi \, d\varphi. \end{aligned} \tag{8.6}$$

Because the $R_{g,i}^2$ are constants, the intensity decay of the unoriented material containing anisotropic particles is

$$\begin{aligned} I(s) = \langle I(\mathbf{s}) \rangle_\omega &= I(0) \left(1 - 4\pi^2 \left(R_{g,1}^2 \langle s_1^2 \rangle_\omega + R_{g,2}^2 \langle s_2^2 \rangle_\omega + R_{g,3}^2 \langle s_3^2 \rangle_\omega\right)\right) \\ &= I(0) \left(1 - 4\pi^2 \langle s_i^2 \rangle_\omega \left(R_{g,1}^2 + R_{g,2}^2 + R_{g,3}^2\right)\right). \end{aligned} \tag{8.7}$$

Finally, comparing with Eq. (8.2) we find

$$\langle s_i^2 \rangle_\omega = \frac{s^2}{3} \tag{8.8}$$

$$R_g^2 = R_{g,1}^2 + R_{g,2}^2 + R_{g,3}^2. \tag{8.9}$$

**Isotropic Particle Scattering: Intensity Decay.** As reported in Eq. (8.2), the isotropic intensity decay is governed by the isotropic radius of gyration, $R_g^2$. The reported relation follows upon solid-angle averaging of the scattering intensity. Starting point for the deduction is the Fourier relation of intensity and sample structure, Eq. (2.14). Equation (8.2) is received after solid-angle average of the harmonic kernel,

$$I(s) = 4\pi \int_0^\infty r^2 \left\langle \Delta \rho^{*2} \right\rangle_\omega (r) \, \frac{\sin(2\pi rs)}{2\pi rs} \, dr,$$

series expansion of $\sin(2\pi rs) / (2\pi rs) = 1 - 2\pi rs/6 + \dots$ and, finally again, the reasoning on the autocorrelation of the particle excess electron density.

## 8.2 Peak Spotting: WAXS Reflections, Long Periods

The evaluation of peaks from scattering patterns (position and shape) is in the focus of the present section. Both isotropic and anisotropic patterns are considered. If the patterns are anisotropic, the anisotropy is not evaluated. Methods for the evaluation of complete scattering patterns will be discussed beginning from Sect. 8.3. Anisotropy is discussed in Chaps. 9 and 10.

### 8.2.1 Discrete and Diffuse Scattering

**What is a peak?** Local intensity maxima or shoulders in scattering patterns are called peaks or reflections, in particular when they are sharp. In USAXS and SAXS even a broad shoulder is called a peak. The set of all peaks is named discrete scattering.

**What is around the peaks?** Before, beneath, and behind the peaks there is diffuse background scattering. The scattering according to GUINIER's law (Sect. 8.1) is an example for such diffuse scattering. Below and after the USAXS, SAXS, and MAXS regions there is diffuse scattering according to POROD's law (Sect. 8.3.2) or diffuse scattering from fractals (Sect. 8.3.3) . In the regime between SAXS and WAXS the diffuse fluctuation background (Sect. 8.3.1) contributes considerably if *soft* condensed matter is studied. The background of the WAXS regime is mainly COMPTON scattering. Figure 8.3 on p. 105 shows both diffuse and discrete scattering components in a curve.

### 8.2.2 Peaks in Isotropic and Anisotropic Scattering Patterns

### 8.2.2.1 Isotropy and Anisotropy

For isotropic materials all reflections represent concentric rings (DEBYE-SCHERRER rings) in an image recorded on 2D detector[5] if during exposure the detector was

---

[5] ... or on its classical analog, plane photographic film

positioned in normal transmission geometry[6] (cf. Fig. 4.1, p. 37). In the center of the rings is the image of the primary beam that, in general, is hidden by a beam stop . In the scattering patterns from anisotropic (i.e., oriented) materials the peaks appear as more or less isolated spots.

### 8.2.2.2 Where to Search for Peaks of Fibers

If the orientation is uniaxial (i.e., fiber symmetrical), the strong peaks of polymer materials are, in general, found in specific regions of the pattern. The *strong WAXS peaks* are found close to the equator[7] of the WAXS pattern. Thus it is good practice to let an offset WAXS detector monitor the equator region.

On the other hand, *strong SAXS reflections* of uniaxial material are, in general, found on the meridian[8]. So be sure that a SAXS detector covers the meridian whenever fiber material or strained polymers are studied.

If expected *narrow* meridional WAXS reflections are not found, the reason may be the bending-away of EWALD'S sphere (cf. Fig. 2.6, p. 28). In this case it may be convenient to tilt the sample with respect to the primary beam (cf. Fig. 2.7, p. 29).

### 8.2.3 WAXS Peaks and Peak Positions

During experiments aiming at WAXS peak analysis, remember to properly calibrate the scattering angle on a 2D detector (Sect. 6.4.1) for setups at a synchrotron beamline or if a rotating anode setup is used. If peak shape shall be discussed later, data quality is an important issue. Do not save exposure time on the cost of signal-to-noise (S/N) ratio. Assess the S/N-ratio from scattering curves, not from gray-scale or pseudo-color images. Images appear much clearer than related curves. Our visual sense is trained to apply a bandpass filter[9].

**Detection of Crystallization and Melting.** A problem frequently tackled by monitoring WAXS peaks is the detection of crystallization and melting as a function of temperature, time, pressure, or other processing parameters → As long as some of the characteristic peaks of a polymer material are observable, a fraction of the material is in crystalline state.

On the other hand, if no peaks are observed, it cannot be concluded that there is no crystallinity. Because the peaks of polymer samples frequently are broad, the

---

[6] If this condition cannot be fulfilled and a detector has to be placed in an offset and tilted position (e.g., for combined SAXS/WAXS measurements), it is the task of beamline staff to provide the user with corrected data or a computer program that compensates the geometric distortion [90].

[7] Reason: In fibers the polymer chains are, in general, oriented parallel to the fiber direction. To a first approximation the chains can be considered an oriented "lattice of rods" with matter and vacuum alternating in equatorial direction. This strong contrast makes strong reflections. On the other hand, regular contrast variations from the chemical structure "along the chain" are much weaker – so the meridional WAXS reflections of polymers are weak, in general.

[8] Reason: Frequently the arrangement of hard and soft domains is more perfect along the principal axis of the oriented material.

[9] Cf. p. 140, and Sect. 8.5.5, where a similar filter is used to extract topological information from scattering data.

typical detection limit is in the order of 1% volume crystallinity [103]. BRAS et al. [104] report that values of 0.1% or even 0.01% may be reached with dedicated low-noise high count-rate detectors at third-generation synchrotron sources.

**Determination of Peak Positions.** Peak positions must be determined for the purpose of crystallographic identification. In isotropic scattering patterns peak positions are the distances of reflections from the origin of the scattering pattern measured in units of $2\theta$, $s$, or $d = 1/s$ – the corresponding Bragg spacing $d_{hkl}$. The triple of indices $hkl$ is called the Miller indices. After identification (indexing) the peak carries three digits (e.g., $d_{200}$). A symbolic short-form for the triple is $h$ – or in this book: $(h)$. Negative digits are indicated by an over-bar, e.g $d_{\bar{1}05}$.

Peak positions may vary[10] as a function of temperature due to thermal expansion or due to conformational changes. Peaks may rest on an inclined background. In this case the background must be subtracted before the peak position is determined, which is simple if the peak is symmetrical.

**Asymmetrical Peaks** are rarely found in WAXS from polymers, but they are ubiquitous in the MAXS of liquid crystalline polymers. For asymmetrical peaks in isotropic patterns it is best to determine the peak position from the maximum of the peak, if peak asymmetry is a result of linear or planar disorder. Linear disorder means that the crystals are more or less one-dimensional (a tower of unit cells). Planar disorder means that the crystallites are made from only very few layers of unit cells (cf. GUINIER [6] Chap. 7).

Microfibrillar structure in isotropic materials makes asymmetrical peaks, because microfibrils are materials with linear disorder. Steep is the increase from small scattering angle. The peak shape can be quantitatively analyzed (STRIBECK [106]) yielding extra information on the lateral extension of the microfibrils.

Asymmetrical peaks with a steep decrease towards high scattering angle are typical for data recorded by a slit-focus (Kratky camera). An isotropic and infinitely sharp peak at $s_0$, $(I(s) = \delta(s - s_0))$, measured by means of an ideal slit becomes $\{I\}_2(s) = J(s) = 2s_0/\sqrt{s_0^2 - s_2^2}$. In practice, the pole at $s_0$ is smoothed from the width of the primary beam.

**Identification of Peaks from Crystallographic Data.** Crystallography is not an issue of X-ray scattering. However, even in materials science crystallographic data are frequently consulted[11]. Based on such data the crystallizing species (component of a blend, block of a block copolymer, one of the crystal modifications possible) can

---

[10]With lower-molecular-weight polymers unit cell parameters may also vary with the molecular mass distribution. For poly(ethylene terephthalate) the history of reported unit cell parameters reflects the progress of chemical processing technology [105].

[11]Find crystallographic data of polymers in: (Polymer Handbook [107], VI-1), (Alexander [7], Appendix 3)

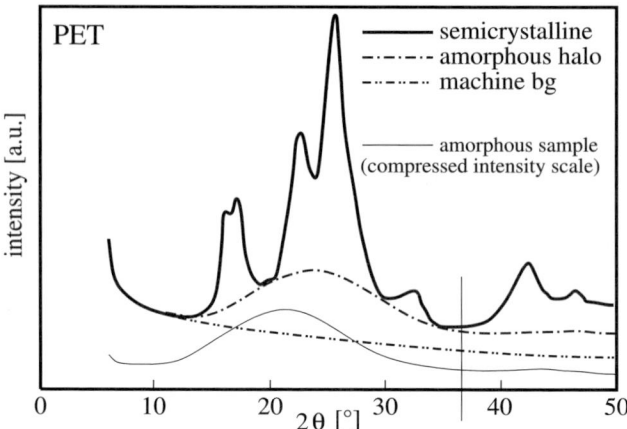

**Figure 8.2.** WAXS curves from semicrystalline and amorphous poly(ethylene terephthalate) (PET). Separation of the observed intensity into crystalline, amorphous, and machine background (laboratory goniometer Philips PW 1078, symmetrical-reflection geometry)

be identified[12]. For common polymers it is frequently more effective to search for a review paper on the requested polymer in order to find crystallographic data and WAXD curves published.

### 8.2.4 Determination of WAXS Crystallinity

In general, it is not recommended to study WAXS crystallinity of anisotropic materials. The recording of the corresponding data is laborious, because the WAXS must be recorded as a function of both scattering angle and sample orientation in a texture goniometer (cf. p. 193, Fig. 9.3) before the data can be isotropized.

#### 8.2.4.1 Phenomenon

For semicrystalline isotropic materials a qualitative measure of crystallinity is directly obtained from the respective WAXS curve. Figure 8.2 demonstrates the phenomenon for poly(ethylene terephthalate) (PET). The curve in bold, solid line shows a WAXS curve with many reflections. The material is a PET with high crystallinity. The thin solid line at the bottom shows a compressed image of the corresponding scattering curve from a completely amorphous sample. Compared to the semicrystalline material it only shows two very broad peaks – the so-called first and second order of the amorphous halo.

It is obvious that the semicrystalline material contains this amorphous feature as well – underneath the reflections. In the semicrystalline material the halo is shifted

---

[12]For this purpose it is helpful to compute a scattering pattern from crystallographic data or vice versa. Freely available is "Powder Cell" by W. KRAUS and G. NOLZE (BAM, 12205 Berlin, Germany). www.bam.de/service/publikationen/powdercell_i.htm

to higher scattering angle, and this is what we expect[13]. In fact, the dash-dotted curve is simply an image of the scattering curve of the amorphous material – affinely stretched both in the vertical and in the horizontal direction.

The dash-dot-dotted curve shows the *machine background* of the goniometer. It is that high, because the machine is set-up in symmetrical-reflection geometry. For a goniometer set-up in normal transmission geometry, the practical background is much less disturbing – in particular at small angles.

Identification of the observed peaks is accomplished by means of data listed in the *Polymer Handbook* [107][14], in reviews or original papers (like the one on PET by GEIL [105]).

### 8.2.4.2 Crystallinity Index

A simple phenomenological method can be used to describe changing crystallinity from WAXS data of isotropic materials. It is based on the computation of areas in Fig. 8.2. First we search the border between first-order and second-order amorphous halo. For PET this is at $2\theta \approx 37°$ (vertical line in the plot). Then we integrate the area between the amorphous halo and the machine background. Let us call the area $I_{am}$. Finally we integrate the area between the crystalline reflections and the amorphous halo, call it $I_{cr}$, and compute a crystallinity *index*

$$X_c = \frac{I_{cr}}{I_{am} + I_{cr}}. \tag{8.10}$$

If several isotropic samples from the same material are studied, $X_c$ arranges them in the order of their crystallinity – but without telling the correct value.

### 8.2.4.3 WAXS Crystallinity for Undistorted Crystals

Undistorted crystals are not found in nature. So the principle discussed here is only a fundament for further reasoning. Every electron in the sample is scattering X-rays by the same amount into some direction in reciprocal space. Thus, the integral of the scattering intensity

$$I_{total} = 4\pi \int_0^\infty s^2 I(s)\, ds$$

taken over the complete reciprocal space[15] is proportional to the total number of electrons in the irradiated volume, i.e., proportional to the sample mass.

If the material is divided in a crystalline and an amorphous phase, and the crystals are undistorted with atoms fixed at their ideal positions in space, the integral

---

[13]The amorphous halo is a result of the fact that there is a preferential distance among chain segments even in an amorphous material. As crystalline layers grow thick, they move the chain entanglements away from the crystal and the entangled amorphous layer becomes compressed. The average distance among the chains is decreased, and in the scattering the maximum position of the corresponding halo is shifted to higher scattering angle (reciprocity in Bragg's law).

[14]If information is not found in the actual issue of the *Polymer Handbook*, it is recommended to consult earlier issues, as well. The focus of the Handbook is changing with time.

[15]As written down here, the equation is valid for isotropic material.

$$I_{crtot} = 4\pi \int_0^\infty s^2 I_{cr}(s)\, ds$$

taken over the crystalline reflections is proportional to the mass of the crystalline phase. In this case the WAXS crystallinity is obtained by

$$w_c = \frac{I_{crtot}}{I_{total}} = \frac{\int_0^\infty s^2 I_{cr}(s)\, ds}{\int_0^\infty s^2 I(s)\, ds}. \tag{8.11}$$

Because we are relating masses to each other, the determined crystallinity is the weight crystallinity $w_c$, not the volume crystallinity, $v_c$.

### 8.2.4.4 WAXS Crystallinity Considering Distortions

Precision determinations of $w_c$ by means of WAXS measurements are carried out by the RULAND [14] method. The method is sufficiently described in textbooks [7, 22]. A modified version adapted to automatic processing by a computer has been introduced by VONK [108].

### 8.2.5 WAXS Line Profile Analysis

### 8.2.5.1 Experimental Technique

Spend *time*[16] on the measurement of the WAXS intensity curve. If noise can be detected by the eye, data are insufficient for further analysis. Correct the raw data for varying absorption as a function of scattering angle depending on the geometry of the beamline setup (Sect. 7.6, p. 76). Measure and eliminate instrumental broadening (cf. Sect. 8.2.5.3). Carry out polarization correction, i.e., divide each intensity by the polarization factor (cf. Sect. 2.2.2 and ( [6], p. 99)). Transform the data to scattering vector representation ($2\theta \to s$). Carry out the LORENTZ correction[17] ($I(s) \to s^2 I(s)$). The resulting curve should look similar to the solid line shown in Fig. 8.3. Then the diffuse background (dotted line in Fig. 8.3) must be defined in such a way that the resulting peaks are symmetrical. Finally the peaks (dashed line in Fig. 8.3) are extracted for further analysis.

### 8.2.5.2 Scientific Goals of Line Profile Analysis

Why are WAXS peaks ("lines") not infinitely sharp?

- The diffractometer cannot produce an infinitely sharp peak (*instrumental broadening*).

- The number of netplanes in the crystal (that are related to an observed peak) is not infinite (*crystal size*).

---

[16]Typical exposure time is 4 - 8 hours using a rotating anode source, Göbel mirror, and a bent 1D detector for simultaneous recording of the complete curve.

[17]Unless the ubiquitous misorientation of crystals in polycrystalline materials has been eliminated by some other method (cf. Chap. 9)

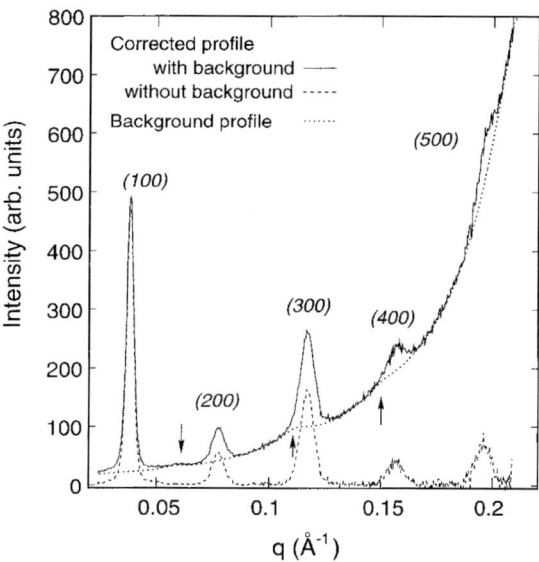

**Figure 8.3.** LORENTZ-polarization corrected WAXS curve of poly(3-dodecylthiophene) before and after background subtraction (from PROSA et al. [109]). The authors define $q$ in the way that is identical to the definition of $s$ in this book

- The crystal itself is not perfect (*lattice distortions*).

Thus line profile analysis is aiming at

- the separation of these effects
- the determination of the kind of lattice distortions
- the quantification of crystal size and lattice distortions

Profound line profile analysis is possible if a *set of reflections* is observed, i.e., peaks from several crystallographic orders indexed by $(h)$ are accessible. With respect to a principal reflection found at the position $s_{(1)}$ the positions of the higher orders are

$$s_{(h)} = (h)\,s_{(1)})$$ (8.12)

according to BRAGG's law. Thus, in a more abstract notion, such a set of peaks $(h)$ is probing physical parameters of the crystal at discrete positions $s_{(h)}$ along a straight radial line in reciprocal space. This is the main idea behind line profile analysis.

To have a set of only two peaks accessible may be sufficient, if additional parameters are varied (temperature [110], pressure [111]).

Both for isotropic and anisotropic scattering patterns line profile analysis can be performed. If a curve from an anisotropic pattern is analyzed, the results are limited to those crystals whose netplanes were oriented in such a way that during measurement they fulfilled BRAGG's reflection condition.

Line breadths are the fundamental quantities in this field of polymer analysis. As a consequence of the Fourier relation between structure and scattering these breadths are *integral breadths*, not "full widths at half-maximum" (FWHM).

### 8.2.5.3 Instrumental Broadening

Consideration of instrumental broadening is a merely technical issue. The instrumental profile $H_I(\mathbf{s})$ must be measured. It is the shape of any peak[18] of a single crystal of "infinite" size and perfection. For application in the field of polymers, many inorganic crystals, e.g., the common standard $LaB_6$, are very good approximations to the ideal case.

The effect of instrumental broadening can be eliminated by deconvolution (see p. 38) of the instrumental profile from the measured spectrum. If deconvolution shall be avoided one can make assumptions on the type[19] of both the instrumental profile and of the remnant line profile. In this case the deconvolution can be carried out analytically, and the result is an algebraic relation between the integral breadths of instrumental and ideal peak profile. From such a relation a linearizing plot can be found (e.g., "measured peak breadths" *vs.* "peak position") in which the instrumental breadth effect can be eliminated (Sect. 8.2.5.8).

### 8.2.5.4 Crystal Size and Lattice Distortion – Separability

Why is it possible to separate crystal size from lattice distortion? — Limited crystal size broadens every reflection by the same amount[20]. On the other hand, the higher the order of a reflection is, the higher is the smearing effect caused by lattice distortions.

Disregarding the crystal size *distribution* in a polycrystalline sample, the observed profile of any peak

$$H_{obs}\left(\mathbf{s} - \mathbf{s}_{(h)}\right) = H_I(\mathbf{s}) \star \Phi_S^2(\mathbf{s}) \star H_D\left(\mathbf{s} - \mathbf{s}_{(h)}\right) \tag{8.13}$$

centered about its position, $\mathbf{s}_{(h)}$, is the convolution of the instrumental profile $H_I(\mathbf{s})$ with the crystal size term[21] $\Phi_S^2(\mathbf{s})$ and the contribution due to lattice distortions, $H_D\left(\mathbf{s} - \mathbf{s}_{(h)}\right)$.

After $H_{obs}$ has properly been extracted (cf. Sect. 2.2.2), the effect of instrumental broadening can be eliminated by numerical deconvolution (see p. 38). If the peaks shall be modeled by analytical functions (Sects. 8.2.5.7-8.2.5.8), the consideration

---

[18] Direct measurement of the primary beam profile may be carried out in the fields of MAXS and SAXS (cf. Sect. 6.4.3).

[19] For instance, Gauss distributions, Lorentz distributions or their combinations

[20] This fact has been discussed in Sect. 2.7.5, p. 24 on the basis of the Fourier convolution theorem (Sect. 2.7.8).

[21] $\mathscr{F}(Y_S(\mathbf{r})) = \Phi_S(\mathbf{s})$, and $Y_S(\mathbf{r})$ is the shape function of the crystal

of instrumental broadening can frequently be deferred and considered in the final linearizing plot.

In the presented form Eq. (8.13) is only valid, if $H_I(\mathbf{s})$ is, indeed, constant over the whole angular range required for analysis. If this is not the case and numerical deconvolution is aimed at, the standard algorithm may be adapted by consideration of the fact that, in any case, the broadening is a slowly varying function of $2\theta$.

### 8.2.5.5 Separation According to WARREN-AVERBACH

A model-free method for the analysis of lattice distortions is readily established from Eq. (8.13). It is an extension of STOKES' [27] method for deconvolution and has been devised by WARREN and AVERBACH [28,29] (textbooks: WARREN [97], Sect. 13.4; GUINIER [6], p. 241-249; ALEXANDER [7], Chap. 7). For the application to common soft matter it is of moderate value only, because the required accuracy of beam profile measurement is rarely achievable. On the other hand, for application to "advanced polymeric materials" its applicability has been demonstrated [109], although the classical graphical method suffers from extensive approximations that reduce its value for the typical polymer with small crystal sizes and stronger distortions.

**Elimination of Instrumental Broadening and Crystal Size Effect.** Fourier transform of Eq. (8.13) turns the convolutions into multiplications (Sect. 2.7.8)

$$h_{obs}(\mathbf{r}) = h_I(\mathbf{r})\, Y_S^{*2}(\mathbf{r})\, h_D(\mathbf{r}). \tag{8.14}$$

If, moreover, we consider a set of peaks with the index $(h)$ counting the orders of reflections, then the effects of size and instrumental broadening are readily eliminated by normalizing

$$\frac{h_{obs(h)}(\mathbf{r})}{h_{obs(i)}(\mathbf{r})} = \frac{h_{D(h)}(\mathbf{r})}{h_{D(i)}(\mathbf{r})} \tag{8.15}$$

the Fourier transform of reflection $(h)$ with respect to that of a different reflection $(i)$ of the set. It is reasonable to "consume" the first peak $((i) = (1))$, i.e., to normalize all subsequent reflections to it.

**Visualizing a Set of Pure Distortion Profiles.** After Fourier back-transformation, we retrieve a set of reduced profiles that are only determined by lattice-distortion

$$H_{D(h-1)}(\mathbf{s}) = \mathscr{F}^{-3}\left(\frac{h_{obs(h)}(\mathbf{r})}{h_{obs(1)}(\mathbf{r})}\right). \tag{8.16}$$

Only one peak has been consumed for normalization purpose. The profiles $H_{D(h-1)}(\mathbf{s})$ contain two important informations, namely

- the peak shape (LORENTZ[22] distribution, Gaussian, ...) that governs the lattice distortions

---

[22]LORENTZ, CAUCHY, or BREIT-WIGNER distribution $h(r) = 1/\left(\pi b\left(1 + (r-r_0)^2/b^2\right)\right)$ at $r_0$ with a full width at half-maximum $b$.

- the law that governs the peak breadth increase with peak order.

**Lattice Distortions of the First Kind.** If the peak breadth is increasing *linearly* with peak order, i.e., if

$$B\left(H_{D(h-1)}(s)\right) \propto (h)$$
$$= \tilde{\sigma}_{LD}(h)\,s_{(1)} \tag{8.17}$$
$$= \tilde{\sigma}_{LD}\,s_{(h)}$$
$$\approx \tilde{\sigma}_{LD}\,s$$

is valid[23], the lattice distortions are called "lattice distortions of the first kind". In lattices subjected to such kind of distortions the long range order among lattice points is preserved. A practical example is strain broadening (see Sect. 8.2.5.6).

The presented variants of Eq. (8.17) may need some explanation. In the second line $\tilde{\sigma}_{LD}$ is introduced as the proportional constant. It is closely related to the relative standard deviation[24] of the distances between the netplanes. The exact relation is established, as soon as a model is introduced. The third line is a result of Eq. (8.12). The last line follows in the limit of a "continuous" set of peaks – an idea that is advantageously applied in several advanced methods for the analysis of highly oriented materials (cf. Chap. 9).

**Lattice Distortions of the Second Kind.** If a *quadratic* increase of the peak breadth with order is found , i.e.,

$$B\left(H_{D(h-1)}(s)\right) \propto (h)^2$$
$$= \left(\tilde{\sigma}_{LD}(h)\,s_{(1)}\right)^2 \tag{8.18}$$
$$= \left(\tilde{\sigma}_{LD}\,s_{(h)}\right)^2$$
$$\approx \tilde{\sigma}_{LD}\,s^2,$$

the lattice distortions are called "lattice distortions of the second kind". In lattices subjected to such kind of distortions there is only short-range correlation among lattice points. An example of such distortions is paracrystalline disorder (cf. Sect. 8.2.5.6).

It is worth to be noted that these definitions of first- and second-order distortions according to WARREN-AVERBACH are model-free. From a linear or a quadratic increase of peak breadths it can neither be concluded in reverse that strain broadening, nor that paracrystalline disorder were detected.

---

[23]Here and in the following $H_{D(h-1)}(s)$ is an abbreviated notation for $\left[H_{D(h-1)}(s)\right]_1(s)$, with $s$ being "the direction in which the scattering curve has been measured in reciprocal space"

[24]Imagine the variation of interplanar distances expressed in percent.

### 8.2.5.6 Matching Lattice Distortions and Structural Models

**Distortions of the First Kind and Thermal Disorder.** In crystallography the best-known example for a lattice distortion of the first kind is the reduction of peak intensity from random temperature movement of the atoms. In materials science a frozen-in thermal disorder of nanostructures[25] is observed as well. The result of this kind of disorder is a multiplicative[26] attenuation of the scattering intensity by the DEBYE-WALLER factor

$$D(s) = \exp\left(-\frac{4}{3}\pi^2 s^2 \langle \Delta^2 u \rangle\right),\qquad(8.19)$$

with $\langle \Delta^2 u \rangle$ designating the mean square deviation of the atom's position from its ideal position[27]. Anisotropy of temperature movement is no issue with polymer materials. Moreover, with polymers the Debye-Waller factor itself is, in general, only a relatively small contribution because strain broadening and paracrystalline disorder are much stronger. GUINIER ([6] Sect. 7.1) presents the deduction and computation of the DEBYE-WALLER factor as a function of temperature for different atoms.

**Distortions of the First Kind and Strain Broadening.** Rolled and drawn metals exhibit a broadening of peaks that increases linearly with the peak order. The first explanation of this observation and its theoretical treatment goes back to KOCHENDÖRFER [112–115]. In the field of SAXS similar considerations have first been published by POROD [18].

The idea is simple: consider a polycrystalline material that is subjected to locally varying strain. Then every crystal is probing its local strain by small compression or expansion of the lattice constant. The superposition of all these *dilated* lattices makes the observable line profiles – and as a function of order their breadth has to increase linearly. According to KOCHENDÖRFER the polycrystalline material becomes "inhomogeneous" or "heterogeneous".

The treatments of KOCHENDÖRFER, POROD, and WARREN-AVERBACH identify "superposition" with the mathematical operation of a convolution. While this is true for translational superposition, for dilational superposition it is a coarse approximation that is only valid for small polydispersity. In the latter case the convolution must be replaced by the Mellin convolution (Eq. (8.85), p. 168): governed by a *dilation factor distribution* and the structure of the reference crystal, the structure of each observed crystal is generated by affine dilation of the reference crystal (STRIBECK [2]).

**Distortions of the Second Kind and Linear Paracrystallinity.** The idea of lattice distortions of the second kind goes back to the famous work of ZERNIKE

---

[25] For instance, inaccurate positions of spherical hard-domains in their lattice of colloidal dimensions

[26] In real space there is a convolution of the ideal atom's position (a delta-function) with the real probability distribution to find it.

[27] Thus $\langle \Delta^2 u \rangle$ is a variance; the 2nd central moment of the probability distribution to find the atom.

and PRINS [116] on liquid scattering: if the distances between atoms are constant the structure is a lattice, but if the distances fluctuate about some average value there is only short-range order. The path from one atom to its $n$-th neighbor is an $n$-fold stochastic process governed by an $n$-fold convolution[28]. This principle makes the line breadths grow quadratically with order. HOSEMANN [5, 117] called the $n$-fold convolution a "convolution polynomial" and a crystal subjected to small crystalline distortions of the second kind a "paracrystal". His mathematical treatment is strict in one dimension (i.e., when applied to *one* set of reflections in WAXS or to a lamellar system in SAXS). In structures of higher dimension it is an approximation only [118], whereas the WARREN-AVERBACH principle presented in Sect. 8.2.5.5 is generally applicable, but does not describe the structure in detail. A proposal of how to combine the WARREN-AVERBACH principle with the paracrystalline model has been presented by BLÖCHL and BONART [119].

### 8.2.5.7 Classical WARREN-AVERBACH Separation

**Overview.** Compared to the Fourier transformation method (cf. Sect. 8.2.5.5), the classical WARREN-AVERBACH evaluation method is a graphical procedure – adapted to low demand of computing power. At least a set of two peaks must be present. The graphical method involves a Guinier approximation (cf. Sect. 8.1) of the function[29] $h_{obs(r)}(h)$ and its evaluation in a Guinier plot. Its results have been proven significant in the field of inorganic materials where the size and distortion effects are small. It is only applicable[30], as long as the peak profiles are symmetrical. A critical examination of the method applied to advanced polymer materials has been published by PROSA et al. [109]. The limits of the approximations made by Warren and Averbach become obvious, if more than two peaks are present and thoroughly analyzed (PROSA et al. [109], Fig. 7[31]). An alternative method with different restrictions has been devised by VAN BERKUM et al . [120].

**Step 1: Measurement and Pre-Evaluation.** For proper data recording and preparation refer to Sect. 8.2.5.1. As a result the peaks (dashed line in Fig. 8.3) are extracted. The curve shows $\sum_{(h)=1}^{5} H_{obs(h)}(s)$ for 5 orders.

**Step 2: Fourier Transform.** Compute the correlation functions $h_{obs(h)}(r) = \Re\left(\mathscr{F}_1\left(H_{obs(h)}\left(s - s_{(h)}\right)\right)\right)$ with $s_{(h)}$ being the center of the peak actually processed. Only few discrete points of each correlation function are actually chosen for the evaluation, namely the points[32] located at

---

[28] ... of a probability distribution to find the next neighbor.

[29] *Notice the change of variables*: GUINIER's law is applied to the Fourier transform of the peak as a function of the *order* $(h)$ *of the peak* at fixed positions $r$.

[30] Moreover: as long as the method is applied to a "scattering intensity curve", i.e., a 1D section in reciprocal space, the analyzed structure is a projection of the correlation function on the respective direction, i.e., an average over planes perpendicular to the direction of the section.

[31] No linear dependence is found in the "Guinier plot" of 5 peaks

[32] Between these points the correlation function should vanish if the lattice were perfect.

$$r \in \left[0, r_{(1)}, 2r_{(1)}, \ldots, 10r_{(1)}\right]$$
$$\in \left[0, L, 2L, \ldots, 10L\right]$$

integer multiples of $r_{(1)} = 1/s_{(1)} := L$, the Bragg spacing of the netplanes related to the studied set of reflections. In the literature $L$ is frequently denoted by $d_{hkl}$. We are only interested in peak shape, so we normalize the correlation function[33]. In original and review literature [28, 29, 109, 120] the resulting values are addressed normalized Fourier coefficients[34]

$$A\left((h), L_j\right) = \frac{h_{obs(h)}\left(jL\right)}{h_{obs(h)}\left(0\right)} = \gamma_{obs(h)}\left(jL\right), \qquad (8.20)$$

and the integer numbers $j$ address the displacement of the crystal with respect to its ghost in steps of size $L$ (causing best matches of netplanes). In fact, the unimaginative presentation of Fourier coefficients $A\left((h), L_j\right)$ is more clearly related to structure after resorting to the correlation function $\gamma_{obs(h)}\left(jL\right)$ with $r = jL$.

### Step 3: Guinier Plot: Separation of Size and Distortion Effects.

The inner part of the correlation functions $\gamma_{obs(h)}\left(r\right)$ is readily expanded into a power series. For this purpose we resort to Eq. (8.14). Assuming that instrumental broadening is already eliminated we have

$$\gamma_{obs(h)}\left(r\right) = \gamma_S\left(r\right)\gamma_{D(h)}\left(r\right) \qquad (8.21)$$

a product of the correlation function of crystal size, $\gamma_S\left(r\right)$, and the correlation function of crystal distortion $\gamma_{D(h)}\left(r\right)$. Let all crystals be infinitely extended in the direction normal to the considered netplanes, all with the same thickness $\ell$. Then the shape correlation is simply $\gamma_S\left(r\right) = 1 - |r|/\ell$ defined for $|r| < \ell$ at the discrete positions $r = jL$. If the crystals are not ideal (finite size and varying shape), the observed shape correlation function becomes a superposition [110]

$$\gamma_S\left(r\right) = \int_0^\infty g_{(h)}\left(\ell\right)\left[1 - \frac{|r|}{\ell}\right] \quad \text{for } |r| < \ell, \qquad (8.22)$$

with $g_{(h)}\left(\ell\right)$ being the number distribution of chords (chord length distribution, CLD) describing the probability to find[35] a column of unit cells of height $\ell$ measured in direction $(h)$.

Let us introduce the number-average chord length $\langle\ell_p\rangle_{(h)}$. For one crystal (of convex shape) it is defined by its volume divided by the shade it throws when illuminated in direction $(h)$. Then a series expansion

---

[33] as is generally done with correlation functions

[34] The term *Fourier coefficient* originates from the theory of Fourier series, in which periodic functions are expanded based on a set of sine- and cosine-functions. The expansion coefficients are called Fourier coefficients.

[35] ... within all the crystals of the polycrystalline sample

$$\gamma_S(r) = 1 - \frac{|r|}{\langle \ell_p \rangle_{(h)}} + \dots \tag{8.23}$$

of the shape correlation function is obtained. Anticipating that WARREN-AVERBACH are aiming at taking the natural logarithm of $\gamma_{obs}$, Eq. (8.23) is substituted into a series expansion of $\log(\gamma_S(r))$ resulting in

$$\log(\gamma_S(r)) \approx -\frac{|r|}{\langle \ell_p \rangle_{(h)}}. \tag{8.24}$$

In similar manner the distortion correlation function $\gamma_{D(h)}(r)$ is treated. Instead of either discussing distortions of the first kind or of the second kind separately,

$$\gamma_{D(h)}(r) = \gamma_{D_1(h)}(r) \, \gamma_{D_2(h)}(r)$$

it is split into two factors[36], which are treated in terms of an approximation for strain broadening and for loss of long range order, respectively. Finally the resulting Taylor series in even powers of $(h)$ are intentionally replaced by Gaussians resulting in

$$\gamma_{obs(h)}(r) \approx \gamma_S(r) \exp\left(-2\pi^2 (h)^2 \left(\frac{r}{L}\right)^2 \frac{\sigma_H^2}{L^2}\right) \exp\left(-2\pi^2 (h)^2 \left(\frac{r}{L}\right)^2 \frac{\sigma_L^2}{L^2}\right) \tag{8.25}$$

with $\sigma_H^2$ the absolute (and $\sigma_H^2/L^2$ the relative) "strain variance". This is the variance of lattice constant variation *from one crystal to another* (first order distortion; heterogeneity of crystals with similar but not identical lattice constants). $\sigma_L$ is the absolute standard deviation of lattice constants *within one crystal* (second order distortion; 1D random variation of lattice constants: "paracrystallinity"). Taking the logarithm of Eq. (8.25) and substituting Eq. (8.24) we finally obtain

$$\log(\gamma_{obs(h)}(r)) = -\frac{|r|}{\langle \ell_p \rangle_{(h)}} - 2\pi^2 (h)^2 \frac{|r|}{L} f_D\left(\frac{|r|}{L}\right) \tag{8.26}$$

an equation simple enough for graphical analysis with the relative variances of both kinds of lattice distortions lumped together in the function

$$f_D\left(\frac{|r|}{L}\right) = \frac{\sigma_L^2}{L^2} + \frac{\sigma_H^2}{L^2} \frac{|r|}{L}. \tag{8.27}$$

Figure 8.4 presents a sketch of the graphical procedure. According to Eq. (8.26) the intercepts in this Guinier plot are

$$\log(\gamma_{obs(0)}(r)) = -\frac{r}{\langle \ell_p \rangle_{(h)}}. \tag{8.28}$$

---

[36]This factorization would be strict only, if (Fourier) convolution were the mathematical operation that describes the effect of both the lattice distortions of the first and the second kind on the profile. In fact, strain broadening is not described by Fourier convolution but by Mellin convolution, instead.

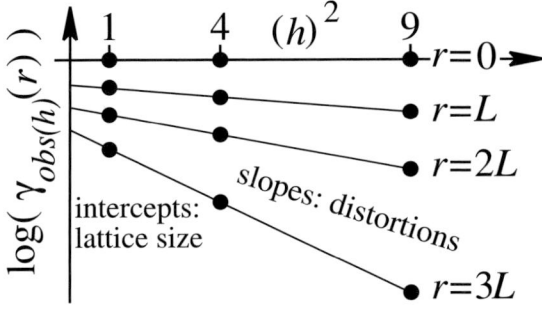

**Figure 8.4.** Graphical separation of lattice size and lattice distortion effects according to WARREN-AVERBACH

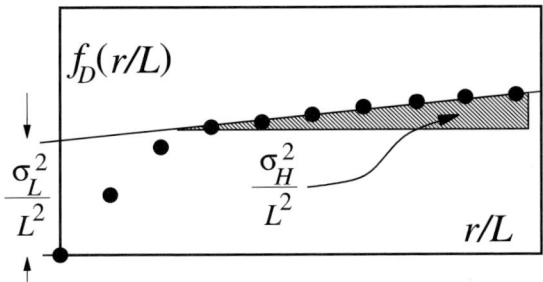

**Figure 8.5.** Graphical separation of lattice distortion effects of the first and the second kind according to WARREN-AVERBACH

In the ideal case they are equidistantly spaced on the ordinate. In reality they may increasingly move closer because real shape correlation functions are flattening as a function of $r$. A significant crystallite size value $\langle \ell_p \rangle_{(h)}$ has only then been determined from this approximative plot, if $r \ll \langle \ell_p \rangle_{(h)}$ was valid for every curve considered in the determination.

As a result of the linear regression, the values of $f_D(r)$ are now known for a sequence of discrete $r$. From Eq. (8.27) it is clear that $f_D(r)$ itself represents a weighted relative variance of the lattice distortions. If it is found to be almost constant, its square root directly describes the total amount of relative lattice distortion ("in percent").

**Step 4: Separation of Distortions of 1st and 2nd Kind.** From Eq. (8.27) the graphical method for the separation of small lattice distortions of the first and the second kind is obvious. It is sketched in Fig. 8.5. In a plot of $f_D(r/L)$ vs. $r/L$ the amount of lattice distortions of the second kind is determined from the intercept. Lattice distortions of the first kind are computed from the slope of the observed curve. The first data points of the curve are, in general, deviating from the straight line and should not be considered for the linear regression [109].

### 8.2.5.8 Separation After Peak Shape Modeling

**Why and How to Use a Model?**   If the graphical method of WARREN-AVERBACH does not work because of weak reflections, overlapping reflections, or problems with the subtraction of background scattering one may resort to modeling the peak shape in Eq. (8.13). Suitable shapes[37] have been resulting from direct peak-shape visualization based on Eq. (8.16) from p. 107. For proper data recording and preparation refer to Sect. 8.2.5.1.

After each peak has been described by the parameters of a model function, the convolution in Eq. (8.13) can be carried out analytically. As a result, equations are obtained that describe the effects of crystal size, lattice distortion, and instrumental broadening[38] on the breadth of the observed peak. Impossible is in this case the separation of different kinds of lattice distortions.

**Polydispersity: Different Crystal Size Averages.**   The crystal sizes in the polycrystalline samples are not identical. So it is important to know, what kind of average (cf. Sect. 1.2) is returned by the method.

The indirect method described here returns the weight-average crystal size [121], irrespective of the model shape chosen. On the other hand, the direct Fourier inversion according to WARREN-AVERBACH returns the number average of the crystal size distribution.

**Model: Gaussian Peaks.**   If all the terms on the right-hand side of Eq. (8.13) can be modeled by Gaussians, the square of the *integral* breadth of the observed peak

$$B^2\left(H_{obs(h)}\left(s\right)\right) = B^2\left(H_I\left(s\right)\right) + B^2\left(\Phi_S^2\left(s\right)\right) + B^2\left(H_{D(h)}\left(s\right)\right) \qquad (8.29)$$

is obtained by summing the squared breadths of the components (WARREN [122], 1941).

**Model: Lorentzian Peaks.**   If all the terms on the right-hand side of Eq. (8.13) can be modeled by LORENTZ curves, the integral breadth of the observed peak

$$B\left(H_{obs(h)}\left(s\right)\right) = B\left(H_I\left(s\right)\right) + B\left(\Phi_S^2\left(s\right)\right) + B\left(H_{D(h)}\left(s\right)\right) \qquad (8.30)$$

is obtained by summing the breadths of the components (HALL [123], 1949).

**Model: Mixed Gaussian and Lorentzian Peaks.**   Even if one of the distributions must be modeled by a Gaussian and the other by a Lorentzian while the instrumental broadening is already eliminated, a solution has been deduced (RULAND [124], 1965).

---

[37]Lorentzians, Gaussians, and combinations of both like pseudo-Voigt functions
[38]Frequently the effect of instrumental broadening is tacitly considered as already eliminated.

**Application.** In practice, three or more peak orders must be observed. The shape of the lines should be Gaussian or Lorentzian. On these premises it is promising to carry out line profile analysis. Corresponding graphical separation methods are readily derived from the breadth relations that have just been discussed, after that the breadths $B\left(\Phi_S^2(s)\right)$ and $B\left(H_{D(h)}(s)\right)$ of a polycrystalline ensemble have been related to structure and substituted in Eq. (8.29) or Eq. (8.30). For the size term the structure relation is

$$B\left(\Phi_S^2(s)\right) = \frac{1}{\langle \ell_p \rangle_{w(h)}}. \tag{8.31}$$

In order to derive this equation, the correlation function in real space is considered, and the Fourier breadth theorem is employed (Sect. 2.7.5, p. 24). $\langle \ell_p \rangle_{w(h)}$ is the *weight* average[39] of the chord lengths of all the crystals in the direction perpendicular to the netplanes of the set of considered reflections.

Resorting to reasoning of KOCHENDÖRFER ([112] and [115], p. 463) an approximation for the breadth of the lattice distortion term due to small amounts of strain broadening

$$B\left(H_{D(h)}(s)\right) \approx 2\,\frac{\sigma_H}{L}\,s_{(h)} \tag{8.32}$$

is obtained by differentiation ($\partial L/\partial s = -1/s^2$) of BRAGG's law ($L = 1/s$) ([112], p. 137) assuming that $\partial L/L$ can be identified with $\sigma_H/L$, because it can practically be considered a "measure of the maximum[40] magnitude of dilation" with respect to the undistorted lattice constant. Moreover, assuming that line broadening is symmetrical about the center of the line[41], Eq. (8.32) is obtained.

On the other hand, lattice distortions of the second kind are considered. Assuming [127] that 1D paracrystalline lattice distortions are described by a Gaussian normal distribution $g_D(\sigma_L, r)$ with standard deviation $\sigma_L$, its Fourier transform $G_D(s) = \exp\left(-2\pi^2\sigma_L^2 s^2\right)$ describes the line broadening in reciprocal space. Utilizing the analytical mathematical relation for the scattering intensity of a 1D paracrystal (cf. Sect. 8.7.3 and [127, 128]), a relation for the integral breadth as a function of the peak position $s_{(h)}$ can be derived [127, 129]

$$B\left(H_{D(h)}(s)\right) = \frac{1}{L}\left(\pi\sigma_L s_{(h)}\right)^2. \tag{8.33}$$

An example for a linearization of integral peak breadths and the corresponding separation of size and distortion effects is sketched in Fig. 8.6. It is clear that at least

---

[39] A short plausible explanation: The series expansion of the graphical WARREN-AVERBACH method is addressing the starting slope of the correlation function – its integral breadth is somewhat wider, if not all the chords are of the same length. The weight-average is obtained.

[40] Instead of considering a maximum magnitude (German: "Höchstwert"), a square average dilation of the lattice constant could have been addressed by Kochendörfer in his approximative reasoning with the same result.

[41] Kochendörfer addresses superposition of "symmetrical lattice line ghosts" originating from the dilation – although it is readily shown that dilatation superposition (Mellin convolution) always causes the observed profile to become asymmetrical [125, 126].

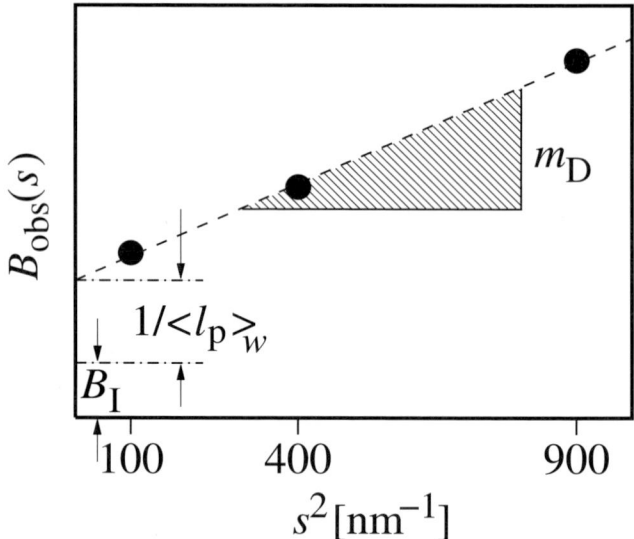

**Figure 8.6.** Separation of lattice distortions and crystallite size according to the WARREN-AVERBACH principle applied to observed integral breadths $B_{obs}(s)$ of a set of WAXS peaks. Because the data are linearized in a plot of $B_{obs}$ vs. $s^2$, the peak shape is (almost) Lorentzian and the predominant lattice distortions are of the second kind with $m_D = \pi^2 \sigma_L^2 / L$. The constant instrumental breadth $B_I$ (determined separately) has not been operated before so that it must be considered now. The remnant intercept is $1/\langle \ell_p \rangle_w$, i.e., the weight-average crystallite size

three lines must have been observed and their integral breadths determined. Then four different plots are tested for the linearization of the peak breadths, and structure parameters are determined from the best linearizing plot. As indicated in Fig. 8.6, there is no need to eliminate instrumental broadening before. It can be considered here, if only the corresponding integral breadth $B_I$ is known and constant. Table 8.1 lists the four different plots and the structural parameters that can be determined from

**Table 8.1.** Integral breadth method according to WARREN-AVERBACH and the four basic possibilities for linearizing plots. All plots are tested for best linearization with the integral breadths from a set of peaks, and the best linearization is taken for structure parameter determination

| Ordinate | Abscissa | Peak shapes | Distortions | Intercept | Slope |
|----------|----------|-------------|-------------|-----------|-------|
| $B_{obs}$ | $s$ | Lorentzian | 1st kind | $B_I + 1/\langle \ell_p \rangle_w$ | $2\,\sigma_H/L$ |
| $B_{obs}$ | $s^2$ | Lorentzian | 2nd kind | $B_I + 1/\langle \ell_p \rangle_w$ | $\pi^2\,\sigma_L^2/L$ |
| $B_{obs}^2$ | $s^2$ | Gaussian | 1st kind | $B_I^2 + 1/\langle \ell_p \rangle_w^2$ | $4\,\sigma_H^2/L^2$ |
| $B_{obs}^2$ | $s^4$ | Gaussian | 2nd kind | $B_I^2 + 1/\langle \ell_p \rangle_w^2$ | $\pi^4\,\sigma_L^4/L^2$ |

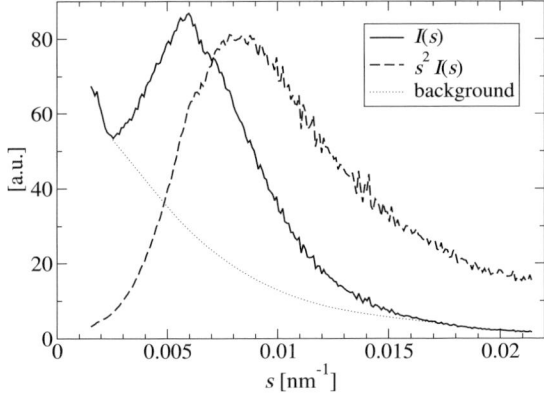

**Figure 8.7.** Peak position determination in SAXS. Isotropic polyethylene during temperature treatment measured at HASYLAB, beamline BW4 (point-focus)

the plot which returns the best linearization of the observed integral peak breadth data.

**Two final remarks.**    It appears worth to be noticed that in the literature the breadth relations are frequently expressed as a function of reflection order $(h) = s_{(h)} L$, and proper variable substitution is an issue. Moreover, the quantity $L$ is frequently identified with edge $a$ of the unit cell. This identification is only valid in a special case [120], as is concluded from simple geometrical reasoning.

### 8.2.6 Peaks in SAXS Patterns

In the most simple analysis of discrete SAXS the observed peak is related to an average distance between nanoscopic domains[42], the *long period*. After determination of the peak position this long period is discussed as a function of materials processing parameters. Modern advanced analysis methods for common polymeric materials follow the advice of DEBYE (cf. p. 1) and consider the complete SAXS pattern to draw conclusions on the short-range correlated structure. SAXS studies in analogy to WAXS methods (crystallographic considerations[43], peak profile analysis[44]) that are less suited for polydisperse materials become important for the study of highly ordered nanostructured materials.

   Peaks in SAXS patterns rest on a rapidly decaying background. Figure 8.7 shows an example for a typical isotropic bulk semicrystalline polymeric material. The long period of such data should never be determined from the peak maximum found in the

---

[42] For instance, crystalline lamellae in an amorphous matrix (semicrystalline polymer materials), hard domains in a soft matrix (thermoplastic elastomers)

[43] If several sharp peaks of colloidal crystals are observed in the SAXS, the unit cell can be determined.

[44] In this case peak profile analysis can be carried out using the methods discussed in Sect. 8.2.5

**Table 8.2.** Long period from the isotropic SAXS data presented in Fig. 8.7 as determined in different ways. The bottom row method is most closely related to reality

| $L$[nm] | determined from |
|---|---|
| 168 | noisy $I(s)$ |
| 170 | smoothed $I(s)$ |
| 127 | noisy $s^2 I(s)$ |
| 121 | smoothed $s^2 I(s)$ |

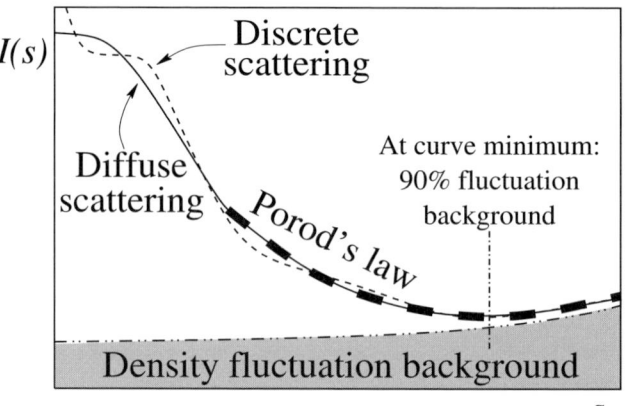

**Figure 8.8.** Important components of diffuse scattering in the SAXS of polymer materials with two or more phases. Only for fractals POROD's law is fundamentally changed

measured SAXS curve, $I(s)$ without background subtraction. It is even better to apply the LORENTZ correction ($I(s) \rightarrow s^2 I(s)$), and noisy curves should be smoothed, before the long period $L = 1/s_L$ is determined from the inverse of the peak maximum position. Table 8.2 shows that different simple ways to determine a long period yield values that differ considerably. Comparison to the advanced methods of correlation function analysis and interface distribution analysis shows that the simple determination from the smoothed and LORENTZ-corrected curve (bottom row in Table 8.2) is quite close to the correct value [130]. Even for moderately anisotropic materials the scattering pattern should be LORENTZ corrected before determination, whereas highly anisotropic materials that show high peaks on a relatively low background do not require LORENTZ correction.

## 8.3 No Peaks: The Interpretation of Diffuse Scattering

Diffuse scattering is always present in the SAXS of polymer materials. In Fig. 8.8 its most important components are sketched[45]. The *density fluctuation background* re-

---

[45] It is assumed that the machine background has separately been measured and properly been subtracted under consideration of the absorption factor (Sect. 7.6). In particular it is not allowed to take the diffuse scattering "background" of a molten or amorphous sample for the machine background.

sults from statistical variations of the electron density within the material. POROD's *law* is found in all polymer materials that are made from a discrete number of distinguishable phases[46]. Only for the rather rare fractal structures it is modified. *Diffuse* scattering of particles is typical for random arrangement of polydisperse particles in a matrix. Although pure particle scattering is most frequently diffuse, it will not be treated here but in Sect. 8.6.

If *polymers* are studied, approximately 90% of the SAXS intensity at the minimum of the curve or the anisotropic pattern is density fluctuation background. For *metals* the corresponding typical value is 10%.

Other effects contribute to the diffuse scattering, as well. In particular, a smooth density transition zone between the phases (e.g., at particle surfaces) and a rough particle surface must be mentioned.

The classical treatment of diffuse SAXS (analysis and elimination) is restricted to isotropic scattering. Separation of its components is frequently impossible or resting on additional assumptions. Anyway, curves have to be manipulated one-by-one in a cumbersome procedure. Discussion of diffuse background can sometimes be avoided if investigations are resorting to time-resolved measurements and subsequent discussion of observed variations of SAXS pattern features. A background elimination procedure that does not require user intervention is based on spatial frequency filtering (cf. p. 140).

### 8.3.1 Intensity Level Between SAXS and WAXS: Electron Density Fluctuations

**Application in Materials Science.** For simple fluids the amount of the density fluctuation background can be computed. Thus its measurement can be used for the calibration of SAXS data to absolute intensity [91, 94]. This method is convenient if liquid samples are studied.

For polymers the density fluctuation background is high. This is a result of its relation to the isothermal compressibility (and the velocity of sound) in the material.

Blending of polymers or the "extension" of polymers with low molecular compounds is changing the density fluctuation background. Thus miscibility can be studied.

As a function of temperature the fluctuation background is changing in a predictable way, and glass transitions in amorphous phases can be studied [93, 95].

**The Phenomenon.** In existing materials the electron density is not even constant inside a single phase. This is obvious for the liquid structure of amorphous regions. Nevertheless, even in crystalline phases lattice distortions and grain boundaries result in variations of the electron density about its mean value. In analogy to the sunlight scattered from the fluctuations of air density, X-rays are scattered from the fluctuations of electron density.

---

[46]For instance: crystalline phase, amorphous phase, hard phase, soft phase, phases formed by different polymeric components in blends or block copolymers.

**Analysis and Significance.**    The study of density fluctuations is part of the scattering theory of amorphous materials. In the experiment amorphous polymers exhibit a slow increase of the fluctuation background as a function of increasing $s$. Only a vague statement is made by existing theory [94, 95, 131–133]: the fluctuation background $I_{Fl}(s)$ is expanded in even powers of $s$. In the *studies of amorphous polymers* the common approximation is

$$I_{Fl}(s) = I_{Fl}(0) \exp\left(\frac{b}{I_{Fl}(0)} s^2\right), \qquad (8.34)$$

or simplified after series expansion

$$I_{Fl}(s) = I_{Fl}(0) + b s^2. \qquad (8.35)$$

Whenever *multiphase polymer materials* are investigated, their peculiar contribution to the diffuse scattering must be considered (POROD's law). The background function becomes more complicated, and with the two-parameter approximation of Eq. (8.34) or Eq. (8.35) the background becomes overparameterized. Thus the common methods of background determination of multiphase polymer materials assume a constant[47] fluctuation background $I_{Fl}$. VONK [134] simply subtracts the minimum of the scattering curve, STEIN [135] and STRIBECK [92] resort to specific plots in which first $I_{Fl}$ can be fixed by the assumption that the correct background is found when a linearized Porod region becomes as long as possible. RULAND and SMARSLY [84] directly fit a topological model for the structure which contains a constant $I_{Fl}$.

    If these concepts of *curve analysis* shall be applied to the anisotropic scattering of polymer fibers, one should choose to study either the longitudinal or the transversal density fluctuations. According to the decision made, the fiber scattering must be projected either on the fiber axis or on the cross-sectional plane. This results in scattering curves with a one- or a two-dimensional PMOROD's law. Because modern radiation sources always feature a point-focus, the required plots for the separation of fluctuation and transition zone are readily established (cf. Table 8.3).

**Density Fluctuations of Fluid Systems and Their X-ray Scattering.**    The theory of fluctuations is a subarea of statistical mechanics and thermodynamics. Respective textbooks are recommended. Here we summarize the frequently cited considerations of RULAND [95, 131, 133]. For a fluid system the probability of a state variable to perform small statistical fluctuations is computed by means of thermodynamics. Here the number of particles (respectively electrons) is of interest. The considerations are simplified by the fact that in scattering experiments the (irradiated) volume is constant and for the *temporal particle density fluctuation*

$$Fl_{N,t} = \langle \rho \rangle \, k_B T \, \kappa_T. \qquad (8.36)$$

---

[47]I.e. constant in the Porod region, e.g., $I_{Fl}(s) \approx const$ for $0.2/\mathrm{nm} < s < 0.35/\mathrm{nm}$, but not for the complete diffuse region $0.2/\mathrm{nm} < s < 1/\mathrm{nm}$ that could be used for an extrapolation according to the procedure used with amorphous polymers.

is obtained[48], which is defined

$$FI_N = \frac{\left\langle (N - \langle N \rangle)^2 \right\rangle}{\langle N \rangle} = \frac{\langle \Delta^2 N \rangle}{\langle N \rangle},$$

by the relative average variance of the number of particles, $N$. Here $\langle N \rangle$ is the average number of particles in the irradiated volume $V$, $\langle \rho \rangle = \langle N \rangle / V$ the average particle density, $k_B$ the Boltzmann constant, $T$ the temperature, and $\kappa_T$ is the *isothermal compressibility* [49] of the material.

If the number $Z_N$ of electrons in each particle is constant, electron density fluctuation

$$FI_{el} = Z_N FI_N \tag{8.37}$$

and particle density fluctuation are proportional. Then the number of electrons

$$N_B(\mathbf{r}_B) = \iiint \rho(\mathbf{r}) Y_B(\mathbf{r} - \mathbf{r}_B) d^3 r$$

in a test volume $Y_B$ placed at $\mathbf{r}_B$ inside the irradiated volume is a convolution. Applying scattering theory and Parseval's theorem RULAND [133] finds

$$FI_{el,v}(Y_B) = \iiint \frac{1}{\rho} \frac{I(\mathbf{s})}{V} \frac{\Phi_B^2(\mathbf{s})}{V_B} d^3 s. \tag{8.38}$$

Here $\rho$ is the electron density of the sample, $I(\mathbf{s})/V$ is the absolute scattering intensity in electron units, $\Phi_B(\mathbf{s}) = \mathscr{F}(Y_B)$ is the Fourier transform of $Y_B$, the shape function of the test volume, and $V_B = \iiint Y_B(\mathbf{r}) d^3 r$ is the size of the test volume.

This equation receives practical importance as the limit to infinite test volume is taken. In this case $\Phi_B(\mathbf{s})$ degenerates to a $\delta$–distribution and

$$\lim_{s \to 0} \frac{1}{\rho} \frac{I(\mathbf{s})}{V} = \lim_{V_B \to \infty} FI_{el,v}(Y_B) \tag{8.39}$$

is obtained. Thus the electron density fluctuation of the sample can be computed from the extrapolation of the measured density fluctuation background to $\mathbf{s} = 0$ if the scattering intensity is calibrated in absolute electron units. For the purpose of extrapolation Eq. (8.35) or Eq. (8.34) is utilized. The choice of suitable linearizing plots is obvious.

### 8.3.2 Intensity Decay Between SAXS and WAXS: POROD's Law

The ideas of the classical evaluation of POROD's law have been developed by VONK [134], RULAND [132, 134], and STEIN [135].

---

[48] SMOLUCHOWSKI (1908), EINSTEIN (1910), ORNSTEIN & ZERNIKE (1914, 1918). In a textbook on scattering HIGGINS & BENOIT ( [136], Sect. 7.6) consider the fluctuation theory from a different point of view.

[49] Thus *soft* condensed matter shows *high* density fluctuation. A relation to the *velocity of sound* is established, if a lattice model is adopted: Expanding an approximation given by GUINIER ([6], Sect. 7.1.6) RULAND concludes [95]: $\lim_{s \to 0} (1/\rho)(I(s)/V) = (\rho/\rho_m)(k_B T)/(v_g^2)$. With $\rho_m$ being the mass density of the sample and $v_g$ the limit of the group velocity of longitudinal lattice waves of long wavelength.

**Basic Equations.** Scattering according to POROD's law [18, 137] is a consequence of phase separation in materials. In a *two-phase system* (e.g., a semicrystalline polymer) every point of the irradiated volume belongs to one of two distinct phases (in the example: to the crystalline phase or to the amorphous phase). In a *multiphase system* there are more than two distinct phases.

Frequently at least one of the phases forms particles (e.g., crystalline lamellae). The shape and position of the $i$th particle from the irradiated volume is described by a shape function $Y_i(\mathbf{r})$. It is obvious that the scattering intensity of an *ideal* multiphase system can be expressed in terms of autocorrelations $Y_i^{*2}(\mathbf{r})$ and cross-correlations of the shape functions and the average electron densities of each phase (cf. Sect. 2.5).

POROD's law reflects the simplicity of the linear series expansion of $Y_i^{*2}(\mathbf{r})$ for *small* $\mathbf{r}$. It says that all multiphase systems exhibit a characteristic decay of the scattering intensity at *large* $s$. The proportionality factor of this decay (POROD's asymptote) is related to the structural parameter of an "average chord length" (cf. p. 112 in Eq. (8.23)). This is the average length that one can travel in the material without crossing a phase boundary. Even the distribution of all chords can be determined from the scattering curve of a multiphase system (*chord length distribution*, CLD).

The scattering patterns from any materials that are made from a finite number of discrete phases conform to POROD's law. In most of the practical applications it is sufficient to consider the most simple case of a *two-phase system*. In-depth considerations concerning the interpretation of scattering data from multiphase systems have been published by JÁNOSI [138].

**Equations.** For a 1D two-phase structure POROD's law is easily deduced. Then the corresponding relations for 2D- and 3D-structures follow from the result. The 1D structure is of practical relevance in the study of fibers [16, 139], because it reflects size and correlation of domains "in fiber direction". Therefore this basic relation is presented here. Let $\mathbf{e_r}$ be[50] the direction of interest (e.g., the fiber direction), then the linear series expansion of the slice $\lceil \gamma(\mathbf{r}) \rceil_{\mathbf{e_r}}$ of the corresponding correlation function is considered. After double derivation the 1D Fourier transform converts the slice into a projection $\{I\}_{\mathbf{e_r}}$ of the scattering intensity and POROD's law

$$\lim_{s \to \infty} s^2 \{I\}_{\mathbf{e_r}} = \frac{k}{2\pi^2 \ell_{px}} := \tilde{A}_{P_1}. \tag{8.40}$$

for the scattering intensity projected on the direction $\mathbf{e_r}$ is established. For large scattering vectors the projected scattering intensity multiplied by $s^2$ is approaching the POROD's *asymptote* for projected scattering data, $\tilde{A}_{P_1}$. $\tilde{A}_{P_1}$ is defined by two material parameters. These are the *invariant* $k$ and the *average chord length* $\ell_{px}$ in the considered direction $\mathbf{e_r}$.

The materials most frequently studied are isotropic. For samples showing isotropic scattering $\ell_{px} = \ell_p$ is valid for any direction. Thus, Eq. (8.40) is generalized resulting in

---

[50]$\mathbf{e_r}$is the vector of length 1 (unit vector) in the chosen direction

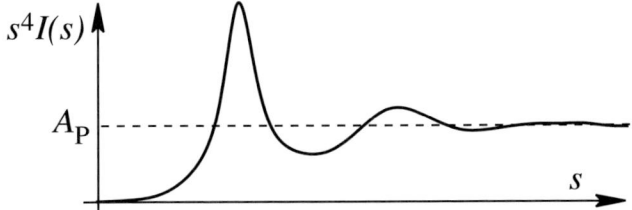

**Figure 8.9.** The scattering curve of an isotropic ideal two-phase system after multiplication by $s^4$ (cf. Eq. (8.43)). The "Porod region" in which the oscillations are almost faded away is generally beginning after the 2nd order of the long period reflection

$$\{I\}_1(s) = \frac{k}{2\pi^2 \ell_p} s^{-2}.$$ (8.41)

For the relation between $\{I\}_1(s)$ and the sought intensity $I(s)$ we have

$$\{I\}_1(s) = 2\pi \int_0^\infty y I\left(\sqrt{s^2 + y^2}\right) dy,$$

and this relation can be inverted after derivation, finally resulting in

$$I(s) = \frac{k}{2\pi^3 \ell_p} s^{-4},$$ (8.42)

$$\lim_{s\to\infty} s^4 I(s) = \frac{k}{2\pi^3 \ell_p} := A_P.$$ (8.43)

for the asymptotic decay of the isotropic scattering intensity of a two-phase system material. After division by the irradiated volume $V$ respective relations for the intensity $I(s)/V$ calibrated in absolute electron units are obtained, in which the practically relevant scattering power (invariant) $Q = k/V$ is replacing $k$. A sketch of the scattering of an isotropic ideal two-phase material is presented in Fig. 8.9.

**The Non-Ideal Two-Phase Structure.**　The real two-phase structure inside a sample is different from the ideal model. The non-ideal two-phase structure is both noisy (cf. Sect. 8.3.1) and blurred. The transformation from a real structure to the ideal one is sketched in Fig. 8.10. The non-ideal system is demonstrated in Fig. 8.10a. The effect of the statistical noise of the electron density is eliminated by subtraction of a fluctuation background, $I_{Fl}$, from the scattering pattern. After that the structure (Fig. 8.10b) is still blurred. A smooth transition zone of thickness $d_z$ between the domains is still present. In the scattering pattern this effect is causing an attenuation of the intensity that is increasing with increasing scattering angle (RULAND [132, 140], VONK [134], STEIN [135]). Compensation of this attenuation results in the scattering curve of the ideal two-phase system.

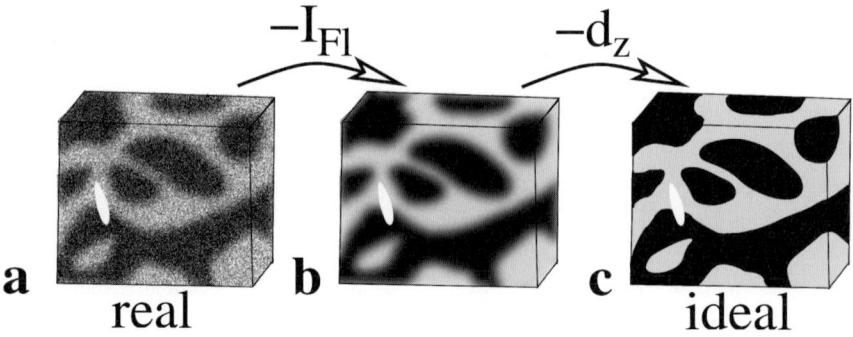

**Figure 8.10.** The real two-phase system (a) and the transition into an ideal system (c) by removal of the density fluctuation background, $I_{Fl}$, and of a transition layer of thickness $d_z$ between the "hard" and the "soft" domains. The elongated white region indicates a void

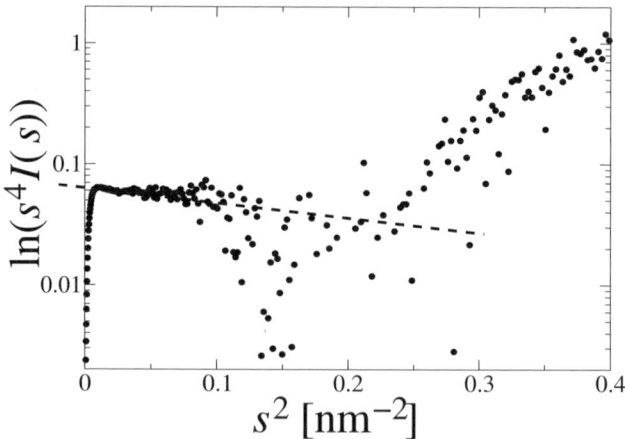

**Figure 8.11.** POROD's law (dashed line) after subtraction of the density fluctuation background $I_{Fl}$ in the scattering curve of an isotropic polyethylene sample measured at a point-focus X-ray beamline

**Classical Evaluation of POROD's Law.**    The practical evaluation of POROD's law in the most common case of an isotropic sample measured with point-focus is demonstrated in Fig. 8.11. By variation of the fluctuation background a long and linear "Porod region" can be received. Nevertheless, the line still shows a negative slope (Fig. 8.11). RULAND's theory of the systematic deviations from POROD's law [132] explains this finding.

> The experimental data presented in Fig. 8.11 show typical noise. Even if the signal-to-noise ratio in the "outskirts" of the scattering curve is improved by experimental technique (e.g., measurement with a 2D de-

**Table 8.3.** Classical Porod-law analysis of three kinds of scattering curves. Linearizing plots and the values of intercept and slope

| Scattering | Symbol | Plot | Intercept | Slope |
|---|---|---|---|---|
| isotropic | $I(s)$ | $\ln\left(s^4 I(s) - I_{Fl}\right)$ vs. $s^2$ | $A_P$ | $\frac{4}{9}\pi^2 d_z^2$ |
| 2D projection | $J(s_{12})$ | $\ln\left(s_{12}^3 J(s) - J_{Fl}\right)$ vs. $s_{12}^2$ | $\frac{\pi}{2}A_P(\mathbf{e}_{12})$ | $\frac{4}{9}\pi^2 d_z^2(\mathbf{e}_{12})$ |
| 1D projection | $\{I\}_1(s_3)$ | $\ln\left(s_3^2\{I\}_1(s_3) - \{I\}_{Fl}\right)$ vs. $s_3^2$ | $\pi A_P(\mathbf{e}_3)$ | $\frac{4}{9}\pi^2 d_z^2(\mathbf{e}_3)$ |

tector and following azimuthal averaging), considerable noise is found in the optimized plot after RULAND [132][51]. The beginning of the wide-angle scattering is, in general, clearly discernible (in the example $s^2 > 0.2\,\mathrm{nm}^{-2}$). In front of the WAXS the Porod region is found. The **apparent asymmetry** of the noise in the swarm of data points is a result of the transformed presentation $\ln\left(s^4\left(I(s) - I_{Fl}\right)\right)$: negative deviations are considerably magnified, whereas positive deviations are demagnified. This distortion has to be considered as $I_{Fl}$ is determined by choosing the value that maximizes the length of POROD's region. From the intercept the Porod asymptote $A_P$ is determined (cf. Fig. 8.9). From the slope of the dashed line the width $d_z$ of the transition zone is computed.

Table 8.3 summarizes the variants of classical Porod-law analysis. For isotropic scattering curves and the most frequent projected curves from anisotropic scattering patterns the plots that linearize the Porod region are indicated together with the quantities that are related to the intercept and the slope of the retrieved line. It is good practice to assess the interval of confidence of the determined parameters. For this purpose the limiting values of the fluctuation background are determined at which the Porod line clearly bends up and down, respectively. Determination of the respective $d_z$ values will give an interval of confidence. Unfortunately for the majority of the analyzed data it turns out that $d_z$ cannot be determined with sufficient accuracy.

To specify the components of the scattering vector by $s_{12}$ and $s_3$ is only a suggestion. The specification meets the case that is of highest practical importance (anisotropy with fiber symmetry).

The 2D projection $\{I\}_2(s_{12}) = J(s_{12})$ in Table 8.3 is denoted by the symbol $J(s)$ – the classical notation of a "slit-smeared" scattering intensity (Kratky camera). Instead of utilizing mathematics, the Kratky camera carries out the 2D projection by means of the engineered slit-focus.

---

[51] As modern one- or two-dimensional detectors are used, every pixel of the detector is enforcedly receiving the same exposure (time). Only by means of an old-fashioned zero-dimensional detector the scattering curve can be scanned in such a manner that every pixel receives the same number of counts with the consequence that the statistical noise is constant at least in a linear plot of the SAXS curve. The cost of this procedure is a recording time of one day per scattering curve.

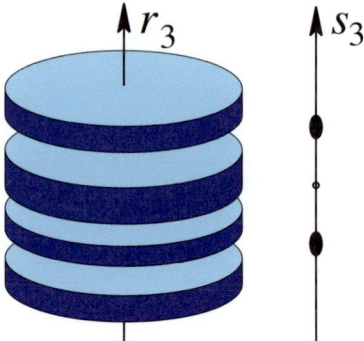

**Figure 8.12.** The structural entity "layer stack with infinite lateral extension" (left) results in a 1D scattering intensity (right)

**Determination of the Average Chord Length.**  The average chord length $\ell_p$ can be determined even from scattering data that are not calibrated. For this purpose $A_P$ is determined from POROD's law in relative units and $k$ is computed by integration of the scattering curve. Finally $\ell_P$ is found from Eq. (8.43).

**Inner Structure of the Porod-Law Exponent.**  In this paragraph it is demonstrated that the Porod-law exponent is the *product* of the Fourier transform of the Laplacian [26] edge enhancement operator $(-4\pi^2 s^2)$ and a solid-angle average. The latter considers misorientation of the scattering entities. If the material is isotropic in three dimensions, this solid-angle average is the LORENTZ factor $(2\pi s^2)$. For materials which show a scattering that is isotropic in two dimensions (slit-smeared intensities $\{I\}_2(s) = J(s)$) the corresponding "LORENTZ factor" is $\pi s$. We will need this relation when it comes to the visualization of nanostructure from scattering data.

Table 8.3 shows that there are Porod laws with exponents $p = 2$, 3 and 4. The exponent $p = 4$ shows up in materials which are isotropic (in 3D space). If we project such a scattering pattern to a plane, the corresponding slit-smeared intensity shows an exponent $p = 3$. The projected scattering pattern is isotropic as well – in the 2D plane onto which it has been projected. Therefore any Porod law has an exponent of at least $p = 2$. The reason is that the scattering of an isotropic ideal multiphase material with sharp edges is readily expressed in terms of the 2nd derivative of its radial correlation function (MÉRING and TCHOUBAR [118, 141]). The derivative theorem yields the factor $-4\pi^2 s^2$ for the scattering intensity, if in real space an isotropic second derivative or a non-isotropic Laplacian is applied (cf. Sect. 2.7.4).

To demonstrate the effect of misorientation or even isotropization let us consider a structural entity[52] which is a perfect lamellar stack. Figure 8.12 demonstrates

---

[52] A *structural entity* is a particle or an ensemble of arranged (i.e., correlated) particles that causes a distinct scattering pattern upon irradiation. Sometimes we call a structural entity made from several particles a *cluster* – not meaning that such particles are touching each other.

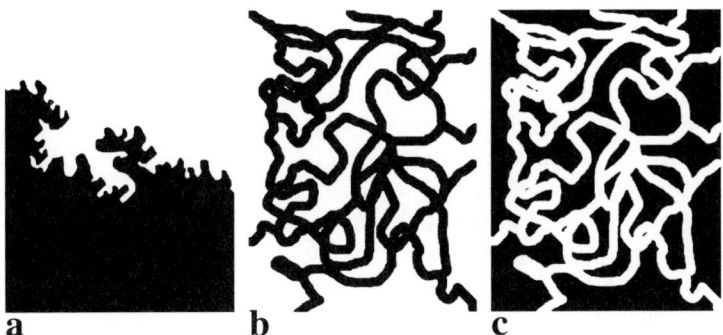

**Figure 8.13.** Different kinds of fractals. (a) Surface fractal, (b) Mass fractal, (c) Pore fractal

the fundamental relation between the structure and the scattering intensity. For a structural entity made from a stack of parallel, infinitely extended layers the related scattering intensity is $I_1(s_3)$. Its restriction to the $s_3$-axis follows from reciprocity (Sect. 2.7.6). Many semicrystalline polymer materials are sufficiently represented by such a layer stack model and its 1D intensity $I_1(s_3)$. If the material is isotropic, $I_1(s_3)$ must be isotropized by solid-angle averaging in 3D, resulting in the observed isotropic intensity, $I(s)$. Reciprocally, the 1D intensity

$$I_1(s) = 2\pi s^2 I(s) \tag{8.44}$$

is obtained from the observed intensity simply by 3D LORENTZ correction. This relation demonstrates how the LORENTZ factor is introduced in the Porod law. Only for layer stacks does this simple relation hold between the isotropic material and the scattering of the perfectly oriented structural entity.

### 8.3.3 SAXS: Fractal Structure

In general we describe structuring of materials by means of *domains*. Frequently such domains are *sufficiently smooth*, and thus surface as well as volume and mass are well-defined parameters. If in Sect. 8.3.2 we would have deduced POROD's law mathematically, we would have handled domain surfaces, shades and the lengths of chords intersecting these domains (e.g., crystalline layers).

What happens, if the surface of the domains becomes increasingly *roughened*? In this case the shape of the diffuse tail of the SAXS is, again, modified. According to RULAND [140, 142] an additional diffuse background is emerging.

Not covered by RULAND's theory is increased roughness exhibiting "foamed" domain surfaces with the pore sizes varying over several orders of magnitude. Such materials are treated in the field of fractals. Because the surface of a solid domain is undergoing fractal roughening, the corresponding fractal is called a surface fractal (Fig. 8.13a). The other well-known type of fractal is the mass fractal in which the domain itself is porous, too. (Fig. 8.13b) Thus in a mass fractal both the solid and its surface exhibit fractal geometry. Mass fractals may be the result of diffusion-limited aggregation. A new kind of fractal structure is the pore fractal [143]. It is the inverse of a mass fractal: the space of solid void are interchanged. Thus in a pore fractal both the pores and their surfaces are fractal (Fig. 8.13c).

Characteristic for a fractal structure is self-similarity. Similar to the mentioned pores that cover "all magnitudes", the general fractal is characterized by the property that typical structuring elements are re-discovered on each scale of magnification. Thus neither the surface of a surface fractal nor volume or surface of a mass fractal can be specified absolutely. We thus leave the application-oriented fundament of materials science. A so-called fractal dimension $D$ becomes the only absolute global parameter of the material.

The *theory* of fractals constitutes $D$ by a *power law*

$$P(\lambda \mathbf{r}) = \lambda^D P(\mathbf{r}), \tag{8.45}$$

which describes how the "property" $P$ of the fractal (e.g., its surface) changes when the characteristic scale $\mathbf{r}$ in the embedding space is dilated by a factor $\lambda$. The fact that $D$ is assumed to be independent of $\mathbf{r}$ is resulting in the abovementioned self-similarity at all scales.

Experimentally accessible is $D$ by means of scattering methods [144]. The corresponding fractal analysis of scattering data is gaining special attractivity from its intriguing simplicity. In a double-logarithmic plot of $I(s)$ *vs.* $s$ the fractal dimension is directly obtained from the slope of the linearized scattering curve. It follows from the theory of fractals that

$$I(s) \propto s^\nu = \begin{cases} s^D & \text{with } 1 < D < 3 \quad \text{for mass fractals} \\ s^{6-D_s} & \text{with } 2 < D_S < 3 \text{ for surface fractals} \end{cases}, \tag{8.46}$$

which means that for the diffuse scattering the observed exponent of decay, $\nu$, is in the range $1 < \nu < 3$ for mass fractals and $3 < \nu < 4$ in the case of surface fractals. The case $\nu = 4$ describes POROD's law (Sect. 8.3.2) and corresponds to the classical Euclidean geometry with domains exhibiting smooth surfaces and sharp density transitions ($D_s = 2$).

Figure 8.14 shows a sketch of the plot that is utilized for the purpose of fractal analysis. For the theoretical fractal self-similarity holds for all orders of magnitude – to be measured in units of space ($r$) or reciprocal space ($s$)[53]. In practice, a fractal regime is limited by a superior cut and a lower cut[54]. In the sketch superior and lower cut limit the fractal region to two orders of magnitude in which self-similarity may be governing the materials structure.

Simply plotting raw diffuse scattering data in a double-logarithmic plot will, most probably result the finding of a linear region [145] that can be interpreted as a fractal. Unsubtracted machine background and fluctuation background will slow the apparent decay of diffuse scattering. Moreover, fractals are possible candidates for

---

[53]It is unreasonable to assess the significance of a fractal structure by resorting to the number of magnitudes covered on the intensity scale.

[54]These quantities are also called inner and outer cutoffs.

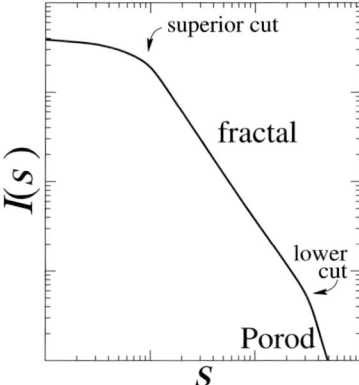

**Figure 8.14.** Sketch of diffuse scattering in the double-logarithmic plot that is used for the determination of the fractal dimension. Superior and lower cut limit the fractal region that should be followed by an interval in which POROD's law is valid

multiple scattering (cf. Sect. 7.2). Thus a SAXS fractal analysis should always be accompanied by other methods that support the interpretation. Specialists in the fields of both fractal analysis and of scattering theory [146, 147] claim that one should be very cautious to label an object a fractal based on simplistic analysis of scattering data in a double-logarithmic plot:

> "The majority of the data that was interpreted in terms of fractality in the surveyed Physical Review journals does not seem to be linked (at least in an obvious way) to existing models and, in fact, does not have theoretical backing. Most of the data represent results from non-equilibrium processes. The common situation is this: An experimentalist performs a resolution analysis and finds a limited-range power law with a value of D smaller than the embedding dimension. Without necessarily resorting to special underlying mechanistic arguments, the experimentalist then often chooses to label the object for which she or he finds this power law a 'fractal'." [146]

Nevertheless, fractal structure is an issue in porous materials.

## 8.4 General Evaluation by Integration of Scattering Data

### 8.4.1 Azimuthal Averaging of Isotropic Scattering Patterns

As 2D detectors are becoming the general standard, even isotropic scattering patterns are frequently measured by means of such an advanced detector. In this case the signal-to-noise ratio of the scattering curve can be increased by azimuthal averaging. Azimuthal averaging means that for each chosen distance from the center of the pattern all the valid pixels on a circular ring are picked. Their intensity is summed (integrated) and divided by the number of the valid pixels. By means of this procedure the averaged scattering curve is obtained.

Application to anisotropic patterns for the purpose of "isotropization" is not reasonable [148]. The operation can directly be performed by *FIT2D*( [39], Chap. 11). If IDL or *pv-wave* shall be used, the library functions DIST() and SHIFT() make the program run fast[55]. If the pedestrian solution of the algorithm is programmed (e.g., in Excel, Java, Pascal, C, ... ) the square-root[56] should not be drawn in order to avoid further reduction of the program speed.

### 8.4.2 Isotropization of Anisotropic Scattering Patterns

Frequent malpractice is the application of the azimuthal averaging algorithm to anisotropic scattering data for the purpose of isotropization. The result appears isotropic, but the chosen integration is incorrect. Only in the case of low anisotropy this procedure is permitted, because then the introduced error is kept small.

If azimuthal averaging is used for the purpose of isotropization, a geometric problem from the 3D world is taken for a 2D problem[57]. For the field of polymer science the correct integration procedure has already been described in 1967 by DESPER and STEIN [148].

In general, only a 2D scattering pattern will be available. In this case isotropization can only be performed if the pattern shows fiber symmetry and the fiber axis is contained in the scattering pattern. This symmetry axis must be known. *Complete* is the available information under these conditions only if SAXS data are evaluated. For WAXS data there are blind regions about the meridian (cf. Fig. 2.6 on p. 28), and missing information must be completed either by extrapolation or by extra experiments in which the sample is tilted with respect to the primary beam.

*Completeness* means that from the recorded data we have to be able to reconstruct the scattering intensity for every point inside a sufficiently big volume of reciprocal space.

**Isotropization in the Case of Fiber Symmetry.** If methods for the analysis of isotropic data shall be applied to scattering patterns with uniaxial orientation, the corresponding isotropic intensity must be computed. By carrying out this integration (the solid-angle average in reciprocal space) the information content of the fiber pattern is reduced. One should consider to apply an analysis of the longitudinal and the transversal structure (cf. Sect. 8.4.3).

We start from a complete map $I(s_{12}, s_3)$ and take into account that in 3D reciprocal space each pixel in this 2D map is representative for many pixels arranged on (half[58]) a ring of iso-intensity. Frequently the sought isotropic analysis method is not

---

[55] As an example download my *pv-wave* programs and consult sf_azimavg.pro

[56] That is, do not compute the distance of a pixel from the center, but only square of the distance. This is sufficient.

[57] Remember Plato's "Allegory of the Cave"

[58] Each pixel represents two full rings, if we only consider the representative quadrant, $I(s_{12}, s_3)$ for $s_{12} > 0$ and $s_3 > 0$ – one above the equator and the other below it.

based on the isotropic scattering intensity $I(s)$, but on the "LORENTZ-corrected" or "one-dimensional" intensity

$$I_1(s) = 2\pi s^2 I(s),$$ (8.47)

and the computation of this quantity is quite simple. Geometric consideration (cf. the following paragraph "Spherical Average of a Fiber Pattern") yields

$$I_1(s) = I_1(s_3) = 2\pi s \int_0^{\pi/2} |s_{12}| I(s_{12}, s_3) \, d\phi.$$ (8.48)

Equation (8.48) is easily converted into an algorithm: each column of the pattern is multiplied by its distance from the fiber axis. After that the azimuthal integration is carried out.

If $I(s)$ instead of $I_1(s)$ is required, Eq. (8.47) is applied. In reality the algorithm is somewhat more complicated[59], because the presence of invalid pixels must be considered. For the solution of such problems methods and peculiar paradigms have been devised in the field of digital image processing (cf. Sect. 2.9).

**Spherical Average of a Fiber Pattern[60].** The spherical average of a general intensity distribution $I(\mathbf{s})$ in reciprocal space is

$$I(s) = \frac{1}{4\pi} \int_{\phi=0}^{2\pi} \int_{\psi=0}^{\pi} I(\mathbf{s}) \sin\phi \, d\phi \, d\psi.$$

The vector $\mathbf{s}$ ($|\mathbf{s}| = s$) can be expressed in Cartesian $(s_1, s_2, s_3)$ or polar $(s, \phi, \psi)$ coordinates with the polar angle[61] $\phi$ and the azimuthal angle[62] $\psi$. The customary symbol $\theta$ should not be used for the polar angle in a treatise on scattering because of a likelihood of confusion with the Bragg angle $\theta$.

In case of a fiber structure, $I(\mathbf{s})$ exhibits cylindrical symmetry $I(\mathbf{s}) = I(s, \phi)$, and because of the central symmetry $I(\mathbf{s}) = I(-\mathbf{s})$ one obtains

$$I(s) = \frac{1}{2} \int_0^{\pi} I(s, \phi) \sin\phi \, d\phi = \int_0^{\pi/2} I(s, \phi) \sin\phi \, d\phi.$$

As an alternative $I(\mathbf{s}) = I(s_{12}, s_3)$ may be expressed in rectangular coordinates with $s_3$ denoting the principal axis and $s_{12} = \sqrt{s_1^2 + s_2^2} = s\cos\phi$ and $s_3 = s\sin\phi$. Thus the isotropized intensity is

$$I(s) = \frac{1}{s} \int_0^{\pi/2} s_{12} I(s_{12}, s_3) \, d\phi,$$ (8.49)

and correspondingly the 1D intensity is

$$I_1(s) = 2\pi s^2 I(s) = 2\pi s \int_0^{\pi/2} s_{12} I(s_{12}, s_3) \, d\phi.$$ (8.50)

---

[59] As an example download my *pv-wave* programs and consult sf_fib2iso1.pro
[60] This paragraph is added on suggestion of W. Ruland
[61] $\phi$ is the angle between $\mathbf{s}$ and the $s_3$-axis
[62] $\psi$ is the angle between the projection of $\mathbf{s}$ on the $(s_1, s_2)$-plane and the $s_1$-axis

### 8.4.3 SAXS Projections

**Motivation.** In this section practical applications of projections in the field of small-angle scattering are devised. Projections are useful to subdivide the complex analysis of a scattering pattern. For example, the invariant separates the non-topological structure parameters from the topology. As the classical method for the determination of the invariant is only working with (isotropic) scattering curves, we make a (one-dimensional) scattering curve from any pattern by projecting it. The resulting curve can be subjected to the classical method. If we have measured a complex fiber pattern, we can separate the topological information on the stacking of the domains in fiber direction from the information that is describing how the domains are arranged in the cross-section of the fiber.

In Sect. 2.7.2 the mathematical definition of a projection has been given. A projection is an operator which maps a multidimensional function on a subspace by means of an integration.

Lately it has become fashionable to compute an arbitrary projection of the scattering data, a so-called "sum of the WAXS" or "sum of the SAXS". Variation of the resulting number is then discussed in terms of structure variation. Such numbers are computed by simply summing the intensity readings from every pixel of the detector. Obviously this number cannot be related to structure and application of this "method" reveals lack of basic analytical skills.

#### 8.4.3.1 Scattering Power (Invariant)

The best-known projection in the field of scattering is the scattering power $k$; it is a number. $k$ is the total scattered intensity[63]

$$k = \{I\}_0 = \int I(\mathbf{s}) \, d^3 s. \tag{8.51}$$

Mathematically spoken $k$ is the zero-dimensional projection $\{I\}_0$ of the scattering intensity. After calibration to absolute units $I(\mathbf{s})$ turns into $I(\mathbf{s})/V$ – its scattering power is known as POROD's invariant

$$Q = k/V. \tag{8.52}$$

In $Q$ the *non-topological structure parameters* of the material's nanostructure are combined. For multiphase systems this fact can be deduced by application of the Fourier-slice theorem and the considerations which lead to POROD's law. In particular, for a two-phase system it follows[64]

---

[63] In many textbooks and papers the scattering power is addressed as the second moment of the isotropic scattering curve. Even though this statement is formally almost correct, it is nevertheless misleading. The multiplication by $2\pi s^2$ that comes into play in the case of isotropic scattering is not related to the formulation of moments of a 1D function. Instead, it is the result of the integration of an isotropic function in 3D space.

[64] The mathematical structure of $Q$ reflects Babinet's theorem, i.e., the fact that the scattering of inverted structures are identical.

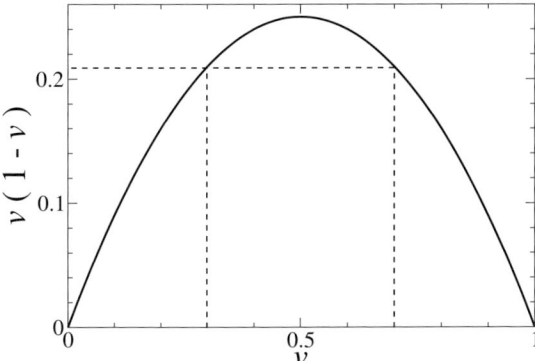

**Figure 8.15.** The invariant as a function of the composition of a two-phase material. Between 30 and 70 vol.-% the scattering power is almost constant. The regions 0 - 30 vol.-% and 70 -100 vol.-% exhibit almost linear relations

$$Q = (\rho_2 - \rho_1)^2 \, v(1-v) \tag{8.53}$$

$$k = (\rho_2 - \rho_1)^2 \, v(1-v) \, V \tag{8.54}$$

with $\rho_1$ and $\rho_2$ being the electron densities of the two phases, and $v$ the volume fraction of one of the phases[65]. $(\rho_2 - \rho_1)$ is the *contrast* between the two phases.

Thus for an ideal two-phase system the total calibrated intensity that is scattered into the reciprocal space is the product of the square of the contrast between the phases and the product of the volume fractions of the phases, $v_1(1-v_1) = v_1 v_2$. $v_1 v_2$ is the *composition parameter*[66] of a two-phase system which is accessible in SAXS experiments. The total intensity of the photons scattered into space is thus independent from the arrangement and the shapes of the particles in the material (i.e., the *topology*). Moreover, Eq. (8.54) shows that "in the raw data" the intensity is as well proportional to the irradiated volume. From this fact a technical procedure to *adjust the intensity* that falls on the detector is readily established. If, for example, we do not receive a number of counts that is sufficient for good counting statistics, we may open the slits or increase the thickness of a thin sample.

If we know that in our two-phase material the volume fraction of one fraction is between 30 and 70%, then the term $v(1-v) \approx 0.23$ is constant to a first approximation (cf. Fig. 8.15). If in this case during the experiment a considerable change of the invariant is observed, it is probably caused by a variation of the contrast[67]. If, on the other hand, the contrast is known to be constant and nanostructure is evolving from a homogeneous phase, the initial increase of the scattering power is proportional to the change of the materials composition.

---

[65] Because $v_1$ and $v_2$ are interchangeable, the index is frequently omitted.

[66] It should be clear that contrast and composition are by no means related to each other. Melting is changing only the composition parameter. Different thermal expansion of crystallites and amorphous matrix is (almost) only changing the contrast.

[67] If we can exclude a change of the irradiated volume by viscous flow of the material.

In many cases the consideration of only *two* phases is sufficient, because according to Eq. (8.53) the scattering power is a function of the *square* of the contrast between two phases. This means that the scattering from a small contrast between a second and a third phase is frequently negligible. For those cases in which three or more phases have to be considered, the corresponding equations have been reported by JÁNOSI [138, 149].

*From experimental* SAXS data of isotropic materials $Q$ is determined by means of Eq. (7.25) (cf. p. 91). An interactive computer program (e.g., TOPAS) is very helpful, because several manual steps are involved.

- First the Porod law and its parameters must be determined (cf. Fig. 8.11, p. 124). We need to know $I_{Fl}$, $A_P$ and the end of PoROD's region (in the example of Fig. 8.11 we have $s_{max} = 0.2$ nm$^{-1}$).

- Second there is a region close to the primary beam where in the integrand[68] no valid data are available. In this region data must be extrapolated. For (old-fashioned) slit-focus cameras this is quite simple, because the corresponding integrand ($\propto sJ(s)$) is increasing linearly from the origin – or, in the case of a lamellar system with one-dimensional density fluctuations from a positive intensity value [70, 150]. For point-focus setup the extrapolation may require a parabolic extrapolation.

- Finally the integration is carried out numerically up to $s_{max}$. The additional term $A_P/s_{max}$ in Eq. (7.25) considers the rest of the integral from $s_{max}$ to infinity. It results from the integration of the *analytical continuation* (Eq. 7.26 on p. 91) of the SAXS intensity by PoROD's law.

The invariant is computed and discussed in many SAXS investigations. There are some studies in which absolute values are determined. Small void fractions in solid materials, the amount of phase separation, or the distribution of plasticizer in a two-phase material can be determined this way. More studies compute relative scattering powers only and monitor materials during processing, most frequently during heating, cooling, or as a function of time after quenching (melting and crystallization).

> In practice, the invariant can be used for the purpose of *calibration to absolute scattering intensity* by means of samples for which the absolute invariant can easily be computed. For this purpose colloidal suspensions of noble metals with known volume concentration are suitable [96]. All the noble metal particles must be small enough so that they really contribute to the observed particle scattering. They must not agglomerate. The raw scattering curve $I_{raw}(s)$ is recorded and reduced to $I_{red}(s)$ by (1) normalizing to the flux of the primary beam, (2) dividing by the thickness of the sample (cuvette) (assuming that the slits will not be changed after this calibration measurement), (3) carrying out the absorption correction. Then the scattering power is computed

---

[68]The integrand is proportional to the dashed curve ($s^2I(s)$) in Fig. 8.7 on p. 117

$$k_{red,noble} = 2\pi \int_0^{s_{max}} s^2 I_{red,noble}(s)\, ds + \frac{A_P}{s_{max}}$$

numerically. Because the electron densities of the solvent (temperature dependence?) and the noble metal are known as well as the *volume*[69] concentration of the noble metal, the invariant $Q_{noble}$ can be computed by means of Eq. (8.53).

Finally, calibration of an unknown scattering pattern is carried out by (1) reducing the intensity in the same way as was done with the scattering of the noble metal sol, (2) obtaining the absolute intensity by

$$\frac{I_{sample}(\mathbf{s})}{V} = \frac{Q_{noble}}{k_{red,noble}} I_{red,sample}(\mathbf{s}).$$

We notice that anisotropic scattering patterns can be calibrated to absolute intensity, as well.

With increasing *temperature* the *contrast* [151, 152] is, in general, increasing, because the thermal expansion coefficient of the soft (amorphous) phase is generally higher than that of the hard (crystalline) phase.

If the scattering power of an *anisotropic material* shall be determined, it is convenient to first project the scattering pattern

$$I(\mathbf{s}) \rightarrow \{I\}_1(s_i)$$

on a line, the direction of which $(s_i)$ may chosen deliberately. Completeness of $I(\mathbf{s})$ is required. This procedure reduces the 3D problem to the evaluation of a scattering curve. The principle of further treatment follows the aforementioned method for isotropic data and Table 8.3. More details are given in the following section.

### 8.4.3.2 1D Projections

While the zero-dimensional projection is only a number, the 1D projection is a curve which can still be evaluated after the projecting integration has been carried out. This means in practice that the evaluation of background and Porod region can be carried out later on the curve $\{I\}_1(s_i)$.

The general definition of a projection has been given on p. 23 in Eq. (2.37). For the purpose of illustration let us write down an example. If $\mathbf{s} = (s_i, s_j, s_k)$ is a representation of the scattering vector in orthogonal Cartesian coordinates, then the aforementioned 1D projection is

$$\{I\}_1(s_i) = \iint_{-\infty}^{\infty} I(s_i, s_j, s_k)\, ds_j\, ds_k. \tag{8.55}$$

Applied to scattering data we encounter the same numerical problems as in the isotropic case: we have to extrapolate inward into the center as well as outward towards infinity. We can avoid the outward extrapolation, if at the outer border of the

---

[69]Remember to convert weight concentration to volume concentration

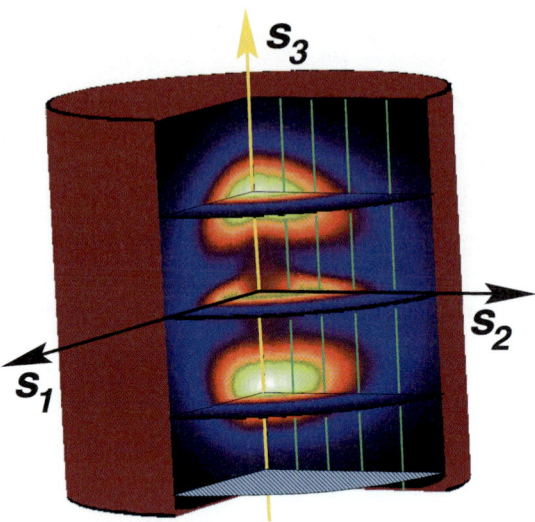

**Figure 8.16.** Demonstration of projections in a fiber diagram. In a projection "on the $s_3$-axis", intensity is integrated over horizontal planes. In a projection "on the $s_{12}$-plane", intensity is integrated along the vertical lines

sensitive detector area the intensity is low enough. The inward extrapolation has to be carried out before starting the numerical integration. For the purpose of such multidimensional extrapolation the radial basis function method has been developed [36]. The corresponding function RADBE is found in the IMSL-library of *pv-wave*.

Particularly useful in materials science is a special 1D projection: the projection of a fiber pattern on the fiber axis, $s_3$

$$\{I\}_1(s_3) = 2\pi \int_0^\infty s_{12} I(s_{12}, s_3) \, ds_{12}. \tag{8.56}$$

In order to demonstrate completeness of a SAXS fiber pattern in the 3D reciprocal space, it is visualized in Fig. 8.16. The sketch shows a recorded 2D SAXS fiber pattern and how it, in fact, fills the reciprocal space by rotation about the fiber axis $s_3$. Let us demonstrate the projection of Eq. (8.56) in the sketch. It is equivalent to, first, integrating horizontal planes in Fig. 8.16 and, second, plotting the computed number at the point where each plane intersects the $s_3$-axis.

Any 1D projection of a SAXS pattern contains specific *one-dimensional* information on the nanostructure[70] of the material. Therefore $\{I\}_1(s_i)$ is called the *one-dimensional scattering intensity* in the direction of $s_i$.

Let us, first, discard the topological information and only determine the scattering power $k$. For this purpose we utilize Table 8.3 and find that the classical

---

[70]For a multiphase structure this information is made from both some direction-dependent topological information, and the non-topological information that is collapsed in the scattering power.

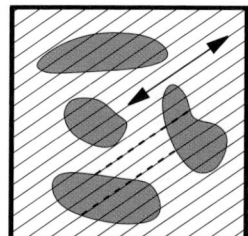 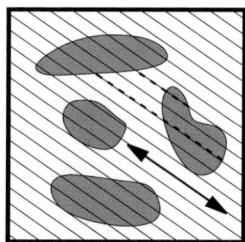

**Figure 8.17.** The topological information on the structure of a multiphase system that is related to one-dimensional projections $\{I\}_1 (s_i)$ in different directions. The demonstration shows two directions indicated by arrows and the related chords. From $\{I\}_1 (s_i)$ the distributions of the chord segments between domain edges are retrieved. Long periods are indicated by broken lines

Porod analysis is carried out in a plot $\ln \left(s_i^2 \{I\}_1 (s_i) - \{I\}_{Fl}\right) vs.\ s_i^2$. We find the number $\{I\}_{Fl}$ by trial-and-error and are satisfied when the linear region becomes longest. We determine the intercept $\tilde{A}_{P_1} = \pi A_P$ and the end of the Porod region, $s_{max}$ (cf. Fig. 8.11). Now we can carry out the numerical integration, again add the remainder term $(\tilde{A}_{P_1} / s_{max})$ from the analytical continuation, and obtain

$$k = 2 \left[ \int_0^{s_{max}} \{I\}_1 (s_i)\, ds_i + \frac{\tilde{A}_{P_1}}{s_{max}} \right] \tag{8.57}$$

for the unnormalized invariant. If we considered projections to different directions, we would find that the parameters $\tilde{A}_{P_1} (\mathbf{e}_i) = \pi A_P (\mathbf{e}_i)$, $s_{max} (\mathbf{e}_i)$ and $\{I\}_{Fl} (\mathbf{e}_i)$ are functions of the projection direction.

The topological information contained in a chosen one-dimensional projection of the scattering intensity is demonstrated in Fig. 8.17. By choosing a direction in the fiber pattern, the respective direction in the materials structure is selected (indicated by a double arrow in Fig. 8.17). Now we imagine a bundle of chords penetrating the multiphase structure in the selected direction. The domain edges cut these chords into segments, and the distributions of all possible segment lengths are generating $\{I\}_1 (s_i)$. Not only single segments are contributing, but also combinations of adjacent segments. Broken lines in the sketch indicate some combined segment lengths that are contributing to the average long period.

Of particular practical value for studies of fibers is $\{I\}_1 (s_3)$ (the intensity projected on the direction of the fiber axis, $s_3$). It measures the average domain extensions and their arrangement in fiber direction. The information contained in this intensity describes the *longitudinal structure* after BONART [16]. In order to demonstrate the chord distributions related to the longitudinal structure we may imagine that one of the directions indicated in Fig. 8.17 were the fiber axis.

From experimental data we would like to *visualize* the segment length distributions – in order both to gain some imagination of the *type*[71] *of the distribution*
– and in order to understand the *arrangement of the domains* in the material. This
visualization is achieved by computation of

- the chord length distribution (CLD) for general isotropic materials (Sect. 8.5.3)

- the interface distribution function (IDF) for 1D projections and the 1D scattering intensity of materials with a structure built from lamellae (Sect. 8.5.4)

- The multidimensional chord distribution function (CDF) for oriented materials
  (in particular useful for the study of materials with uniaxial orientation, i.e.,
  fibers) (Sect. 8.5.5)

We notice that the 1D projections are perfect candidates for structure modeling by
1D models: arrange sticks in a row! For this purpose define stick-length distributions
and the law of their arrangement. Fit such models to the measured scattering data
(Sect. 8.7).

### 8.4.3.3 2D Projections

A well-known device that performs a 2D projection of the scattering pattern is the
Kratky camera. By integrating the intensity along the direction of the focus slit, it
is collapsing the SAXS intensity on the plane that is normal to the slit direction. In
general, 2D projections collapse the measured complete intensity not on a line, but on
a plane. As in the case of the 1D projections, the orientation of this plane can freely
be chosen. The result of such a projection $\{I\}_2(s_j, s_k)$ is not a curve as was the case
with the 1D projection, but a 2D scattering pattern. Only in the case of 2D isotropy
(i.e., $\{I\}_2(s_{jk})$ with $s_{jk} = \sqrt{s_j^2 + s_k^2}$) the scattering pattern can be *represented* by a
curve.

Such 2D isotropy is fulfilled in the case which is of the highest practical value.
Here the 2D projection

$$\{I\}_2(s_{12}) = 2 \int_0^\infty I(s_{12}, s_3) \, ds_3 \tag{8.58}$$

describes BONART's [16] transversal structure of a fiber – the arrangement of domain
cross-sections in the fiber cross-section. Figure 8.18 demonstrates the structure. In
analogy to the 1D projections, chords can be imagined to penetrate the representative
cross-section "in the plane". They become segmented by the circular domain cross-
sections. Finally the segment length distributions generate $\{I\}_2(s_{12})$.

## 8.5 Visualization of Domain Topology from SAXS Data

After we have discussed the composition parameters of the SAXS of a multiphase
material, we now start with the investigation of the topology. The most simple access to the arrangement of domains in the material is the discussion of long period

---

[71] Are the distributions Gaussians, Lorentzians, or even more complex?

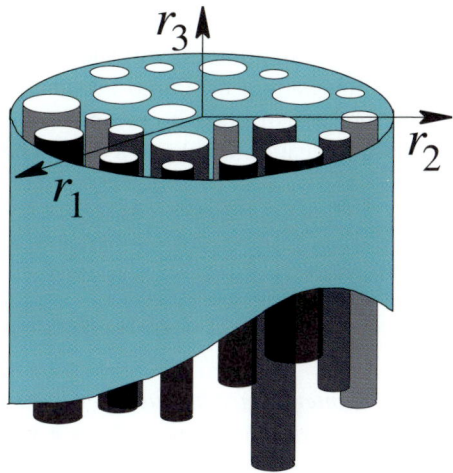

**Figure 8.18.** Transversal structure of a fiber. The topological information on the structure of a fiber that is related to the 2D projection $\{I\}_2(s_{12})$ contains structure information from the representative cross-sectional plane $(r_1, r_2)$ of the fiber. Size distribution and arrangement of the domain cross-sections are revealed

peaks (cf. Sect. 8.2.6). The next level of analysis is visualization of topology. Only for nearly monodisperse or highly oriented materials we should skip this step and directly proceed to a modeling of the structure and fitting of the scattering data[72]. As we have just learned, topology information is only a part of the information buried in a SAXS pattern. So before topology can be visualized, the respective information must be extracted from the scattering pattern.

### 8.5.1 Extraction of the Topological Information

For the scattering of an isotropic material we already know the result of the separation and a method to obtain it: the result is the scattering of the ideal multiphase system as sketched on p. 123 in Fig. 8.9. A way to obtain the result is the classical Porod-law analysis (Sect. 8.3.2).

The fundamental problem of the classical method is the fact that there is no viable[73] procedure to extend it to the scattering of anisotropic materials. Moreover, the required manual processing is cumbersome, slow and may yield biased results.

**The Interference Function.** The function sketched in Fig. 8.9 can be understood as $s^p I_{id}(s)$, the intensity of the *ideal* multiphase system multiplied by a power

---

[72]The background for this advice is explained in the discussion of Fig. 8.35, p. 162.

[73]Conceded – we can successively project anisotropic data to different directions, carry out the manual procedure for each direction, and, finally recombine the curves. But even if we replace the manual procedure by an automated one, the combination of curves turns out to be not contiguous: There is no smooth surface anymore.

$p$ of the modulus of the scattering vector. In the example $p = 4$ compensates the decay of the Porod law. The scattering intensity of the ideal multiphase system is readily obtained by dividing the result of the Porod-law analysis by $s^p$. $I_{id}(s)$ is the starting point for nanostructure visualization by means of the correlation function (cf. Sect. 8.5.2). By a small modification we obtain a well-behaved function

$$G_1(s) = s^p I_{id}(s) - A_p, \qquad (8.59)$$

as it is vanishing in the limit

$$\lim_{s \to \infty} G_1(s) = 0.$$

Equation (8.59) defines the 1D *interference function* of a layer stack material. $G_1(s)$ is one-dimensional, because $p$ has been chosen in such a way that it extinguishes the decay of the Porod law. Its application is restricted to a layer system, because misorientation has been extinguished by LORENTZ correction. If the intensity were isotropic but the scattering entities were no layer stacks, one would first project the isotropic intensity on a line and then proceed with a Porod analysis based on $p = 2$. For the computation of multidimensional anisotropic interference functions one would choose $p = 2$ in any case, and misorientation would be kept in the state as it is found. If one did not intend to keep the state of misorientation, one would first desmear the anisotropic scattering data from the orientation distribution of the scattering entities (Sect. 9.7).

The addressed types of interference functions are the starting point for the evaluations described in Sects. 8.5.3-8.5.5.

**Automated Extraction of Interference Functions.** For the classical synthetic polymer materials it is, in general, possible to strip the interference function from the scattering data by an algorithm that does not require user intervention. Quantitative information on the non-topological parameters is lost (STRIBECK [26, 153]). The method is particularly useful if extensive data sets from time-resolved experiments of nanostructure evolution must be processed. Background ideas and references are presented in the sequel.

Concerning the notions on the deviations of the real structure from an ideal multiphase topology, a survey shows that all models are resulting in *slowly varying* backgrounds of the scattering pattern. On the other hand, noise originating from counting statistics is displaying *high-frequency* deviations of the measured signal from the smooth shape of the scattering. If we are investigating polydisperse soft materials, the observed reflections are broad, i.e., they do not contain high spatial frequencies. Under these conditions the extraction of the topological information can be considered a problem of *signal processing*. The power spectrum of the measured SAXS data shows three distinct bands: backgrounds are in the low spatial frequencies, in the high spatial frequencies there is only noise (because of lacking long-range order) – and the spatial frequency band of polydisperse topology is in between. Thus background can be removed from the scattering pattern by spatial frequency filtering

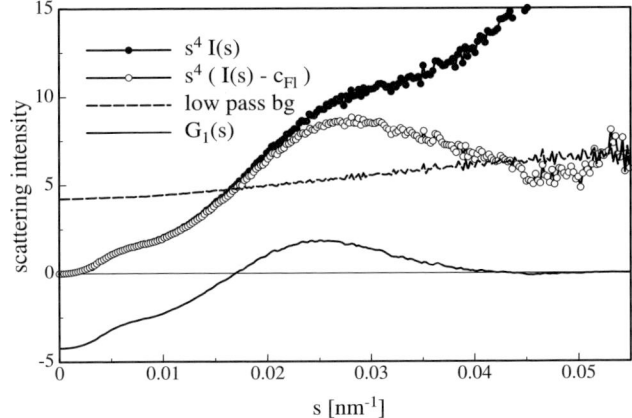

**Figure 8.19.** Extraction of the scattering of an ideal two-phase structure from the raw scattering data of an isotropic UHMWPE material by means of spatial frequency filtering

returning an interference function $G_f(\mathbf{s}) \approx mG(\mathbf{s})$ with $m < 1$ being an unknown factor that describes some loss in the filter, and $G(\mathbf{s})$ the true interference function.

We observe that this method works for data of any dimensionality. Figure 8.19 demonstrates the extraction of the interference function for the case of an isotropic ultra-high molecular-weight polyethylene material. From the raw data (filled circles) a constant fluctuation background estimate[74] $c_{Fl}$ is subtracted in order to ease the task of the filter. The background obtained by low-pass filtering is the dashed line. After subtraction of the background and multiplication by the "Cosine-bell function" (Hanning filter) [26, 153, 154] the interference function (solid line) is received[75].

It is clear that this procedure can be iterated. Iteration successively improves the "balance" of the interference function – and theory says that $\int G(s)\,ds = 0$ should be perfectly balanced if the domain surfaces (e.g., the surfaces of the crystalline lamellae) are smooth. Thus we can interpret iterative spatial frequency filtering as a method to remove the effect of a rough phase boundary. Inevitably this goes along with the extinction of the scattering effect of small domains (e.g., small crystallites). Therefore, removing roughness by iterative spatial frequency filtering is only a last resort for those few materials with very rough [155] domain boundary.

---

[74] See p. 118, Fig. 8.8 and the corresponding comparison of soft matter and metals: Subtract 90% of the intensity minimum.

[75] Assistance of how to choose and to write the low pass filter, the use of the Hanning filter, etc., can be found in textbooks of digital image processing or in the "Reference Guide" of *pv-wave*. A command (#FILTINT) to carry out spatial frequency filtering of scattering curves is part of my program TOPAS. If fiber patterns shall be evaluated, a *pv-wave* procedure sf_interfer does the main job. Both examples are available as source codes (cf. p. 29).

### 8.5.2 1D Correlation Function Analysis

The easiest way to get some impression of the structure behind our scattering data without resorting to models is the computation and interpretation of a correlation function. We will mainly discuss the 1D correlation function, $\gamma_1(r_3)$, because any slice of an anisotropic correlation function is a one-dimensional correlation function. Moreover, $\gamma_1(r_3)$, is readily describing the topology of certain frequent structural entities (stacks made from layers and microfibrils). There is an advantage of the correlation function analysis as compared to "long period interpretation". The analysis of the correlation function permits to determine the *average domain thicknesses* (for example the thicknesses of crystalline and amorphous layers). The principal disadvantage of the correlation function is the fact that polydispersity is not properly reflected in the correlation function [2]. This means that the statistics of domain thickness variation is very difficult to study from a correlation function. In particular, for the latter purpose it is more appropriate to carry out an analysis of the IDF or of the CDF.

**1D Structural Entities.**    In materials science, structural entities which can satisfactorily be represented by layer stacks are ubiquitous. In the field of polymers they have been known for a long time [156]. Similar is the microfibrillar [157] structure. Compared to the microfibrils, the layer stacks are distinguished by the large lateral extension of their constituting domains. Both entities share the property that their two-phase structure is predominantly described by a 1D density function, $\Delta\rho(r_3)$, which is varying along the principal axis, $r_3$, of the structural entity.

**1D Intensity.**    As already mentioned (cf. p. 126 and Fig. 8.12), the isotropic scattering of a layer-stack structure is easily "desmeared" from the random orientation of its entities by LORENTZ correction (Eq. 8.44). For materials with microfibrillar structure this is more difficult. Fortunately microfibrils are, in general, found in highly oriented fiber materials where they are oriented in fiber direction. In this case the one-dimensional intensity in fiber direction,

$$I_1(s_3) = 2\pi \int_0^\infty s_{12} I(s_{12}, s_3)\, ds_{12},$$

can directly be interpreted as the 1D intensity of the microfibrils along their principal axis. By projecting we loose the information on the thickness of the microfibrillar strands and on possible lateral correlations among them.

> **Warning.**    For isotropic materials the 1D projection $\{I\}_1$ and the LORENTZ correction yield different 1D intensities. Both are related by
>
> $$I_1(s) = 2\pi s^2 I(s) = -s\frac{d}{ds}[\{I\}_1(s)]. \qquad (8.60)$$

Model functions for the 1D intensity have early been developed [128, 158] and fitted to scattering data. The classical model-free structure visualization goes back to

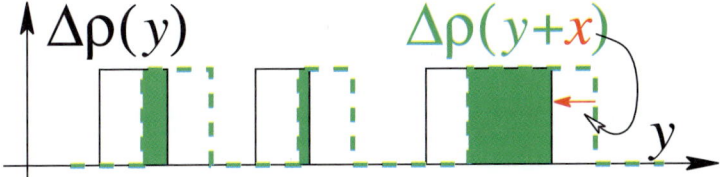

**Figure 8.20.** Generation of a 1D correlation function, $\gamma_1(x)$, by autocorrelation of the 1D electron density, $\Delta\rho(y)$ for a two-phase topology. Each value of $\gamma_1(x)$ is proportional to the overlap integral (total shaded area) of the density and its displaced ghost

VONK [159, 160] and describes the structure by the 1D correlation function $\gamma_1(r_3)$ in physical space.

**Computation of a 1D Correlation Function.** Each *one*-dimensional correlation function, $\lceil\gamma\rceil_1(x)$ or $\gamma_1(x)$ (with $x = r_3$)

$$\lceil\gamma\rceil_1(x) = \frac{2}{k}\int_0^\infty \{I\}_1(s)\cos(2\pi xs)\,ds \qquad (8.61)$$

$$\gamma_1(x) = \frac{2}{k}\int_0^\infty I_1(s)\cos(2\pi xs)\,ds \qquad (8.62)$$

$$= \frac{4\pi}{k}\int_0^\infty s^2 I(s)\cos(2\pi xs)\,ds \qquad (8.63)$$

is computed from its 1D intensity, $\{I\}_1(s)$ or $I_1(s)$, by a *one*-dimensional Fourier transform. Equation (8.63) is valid[76] for the isotropic scattering of a lamellar multi-phase system.

Numerically the correlation function is easily computed, after either a classical POROD-law analysis has been carried out (Sect. 8.3.2), or the interference function has been obtained by spatial frequency filtering (p. 140). For the purpose of extending the integral we may write an adapted Fourier-transformation algorithm which explicitly utilizes the analytical continuation according to POROD's law, or we may use the continuation for the generation of additional grid points and employ the discrete fast Fourier transformation (DFFT) algorithm [154, 161].

Figure 8.20 demonstrates the generation of $\gamma_1(x)$ by displacement of the 1D electron density[77], $\Delta\rho(y)$, with respect to its ghost, $\Delta\rho(y+x)$, along the stack axis $y$. The direction $x$ in the sketch is identical to the direction of the stack normal, $r_3$, in Fig. 8.12 on p. 126. The sketch depicts a displacement $x$ that is still so small that each domain is only correlated to itself (shaded areas). At such a position $x$ we are still in

---

[76]The presented result is different from the *radial correlation function* $\gamma(r) = 4\pi\int s^2 I(s)(\sin(2\pi rs)/2\pi rs)\,ds$, which is computed from the isotropic scattering intensity by means of the *three*-dimensional Fourier transform.

[77]For the sake of simplified presentation here, it is assumed that there are only few scattering entities in a sea of matrix material, and the average $\langle\rho\rangle_V \approx \rho_1$ is close to the density of the matrix phase

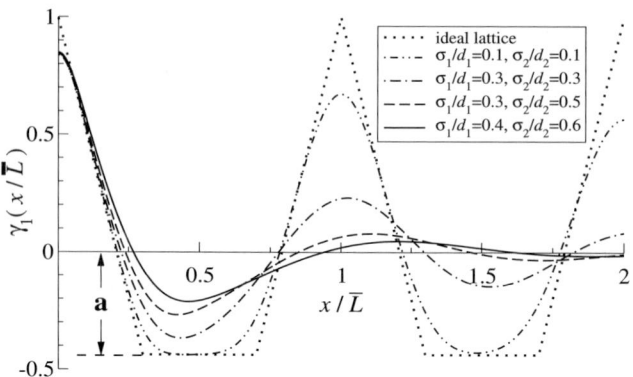

**Figure 8.21.** Features of a 1D correlation function, $\gamma_1\left(x/\bar{L}\right)$ for perfect and disordered topologies. $\bar{L}$ is the number-average distance of the domains from each other (i.e., long period). *Dotted:* Perfect lattice. *Dashed and solid lines:* Paracrystalline stacks with increasing disorder. $a = -v_{l_1}/\left(1 - v_{l_1}\right)$ with $0 < v_{l_1} \leq 0.5$ is a measure of the linear volume "crystallinity" in the material, which is either $v_{l_1}$ or $1 - v_{l_1}$

the region of linear decay of the correlation function, i.e., in the so-called "autocorrelation triangle". The typical shape of such a correlation function for topologies with varying amount of disorder is sketched in Fig. 8.21. Obviously, the autocorrelation triangle of the ideal lattice (dotted curve) is not preserved in paracrystalline stacks of higher polydispersity. Thus, a simple linear extrapolation ("linear regression autocorrelation triangle", LRAT [162]) will only yield reliable information concerning the properties of the idealized lattice from the real data, if the polydispersity remains rather low.

**Analysis of the 1D Correlation Function.**    Several publications describe the search for a simple *graphical* analysis [22, 159, 162–164] of the 1D correlation function by means of a geometrical construction. It is the drawback of all such methods that polydispersity and heterogeneity are not considered. The methods are derived from the general generation principle of correlation functions (Fig. 8.20), resulting in equations (cf. Eqs. (8.23), (8.70) and (8.64)) for the first off-origin maximum, the depth of the first minimum or the *initial* slope $\gamma'_{id}(0)$ of ideal correlation functions. For the simplified case of a lamellar system we obtain

$$\gamma_{1,id}(x) = 1 - \frac{1}{\ell_{p1}}|x| + \dots \tag{8.64}$$

with $\ell_{p1}$ being the average chord length of the one-dimensional ideal two-phase topology with

$$\frac{1}{\ell_{p1}} = \frac{1}{\bar{d}_1} + \frac{1}{\bar{d}_2}, \tag{8.65}$$

$\bar{d}_1$ the average layer thickness of the first of the two kinds[78] of lamellae, and $\bar{d}_2$ related to the second kind of layers. $\bar{L} = \bar{d}_1 + \bar{d}_2$ is called the average[79] long period. Without loss of generality we may restrict further discussion to *linear crystallinities*[80]

$$v_l = \frac{\bar{d}_1}{\bar{L}} \qquad (8.66)$$

with $v_l \leq 0.5$. The crystallinity is called "linear" in order to distinguish it from the overall volume crystallinity in the sample, because $v_l$ does not account for the presence of extended domains (of matrix material) outside the scattering entities. From Eqs. (8.64) and (8.65) we obtain for the zero of the initial slope of the ideal correlation function

$$x_0 = \frac{\bar{d}_1 \bar{d}_2}{\bar{d}_1 + \bar{d}_2}$$
$$= v_l (1 - v_l) \bar{L}. \qquad (8.67)$$

Figure 8.21 shows model functions both for ideal and realistic cases. The dotted curve demonstrates the case of the ideal and infinitely extended 1D lattice. Here every time the ghost is displaced by an integer multiple of the lattice constant ($x/L = 1, 2, 3, \ldots$), the correlation returns to the ideal value 1. For the 1D lattice not only $x_0$, but also the valley depths

$$\gamma_{1,min} = a = -\frac{v_l}{1 - v_l} \qquad (8.68)$$

are related to the composition[81], $v_l (1 - v_l)$, of the material (see also p. 133, Fig. 8.15). The common graphical evaluation methods try to transfer these features of the ideal correlation function of an ideal lattice to real correlation functions of polydisperse soft matter that are computed from experimental data. The valley-depth method has first been devised by VONK [159]: whenever a flat minimum is found in a real correlation function, the distortion is weak and the linear crystallinity can significantly be determined from the properly normalized correlation function by application of Eq. (8.68).

In practice, the observed distortion is frequently strong. Thus, the correlation-function minimum is not flat. This is demonstrated in most of the dashed and solid curves in Fig. 8.21. They show model correlation functions of the paracrystalline stacking model with varying amount of disorder. Computation[82] is based on Eq. (8.104), p. 180.

---

[78]For instance the "amorphous", "hard", "crystalline", …

[79]Speaking of averages and denoting symbols by an overbar already means a generalization for distorted structures which will be discussed later.

[80]Again, "crystallinity" may be replaced by "hard phase fraction", "soft phase fraction", or whatever designation applies better to the material that is studied.

[81]Conceded – Eq. (8.68) violates Babinet's theorem. Nevertheless, it is valid for $v_1 \leq 0.5$ and can easily be remembered, whereas the correct equation is somewhat more involved.

[82]It is convenient to set $A_{P_1} = 1$, $\bar{L} = \bar{d}_1 + \bar{d}_2 = 1$. Rounding errors are suppressed by replacing the intensity by $1/s^2$ (POROD's law) for big arguments ($s > 8$). A smooth phase transition zone (in all the example curves: $d_z = 0.1$) is considered by multiplication with $\exp\left(-(2\pi s d_z/3)^2\right)$. From this one-dimensional scattering intensity the correlation function is obtained by Fourier transformation.

Figure 8.21 shows functions of the distorted topologies that are *not pointed at the origin*, and $\gamma_1(0) < 1$. The reason is that the presented model is not an ideal two-phase system, because it considers smooth transitions of the electron density between the "crystalline" and the "amorphous" layers.

> In practice, even a *more severe damping* of the correlation function close to the origin *is frequently accepted* in order to compute the correlation function with little effort of evaluation [159]: POROD's law is not evaluated (cf. p. 124, Fig. 8.11), and thus the Fourier integral cannot be extended to infinity. Instead, the position $s_{min}$ in the scattering curve is determined at which the SAXS intensity is lowest. This level is subtracted, and the integral is only extended up to $s_{min}$.

The case of low distortion is shown in the dashed-dotted-dotted curve from Fig. 8.21. The first minimum still reaches the ideal valley depth. Therefore it is still possible to determine the linear composition of the material from Eq. (8.68).

Let us discuss the first off-origin maximum of $\gamma_1(x/\bar{L})$. For the ideal lattice and weakly distorted materials the maximum is found at the position of the number-average long period, $\bar{L}$, i.e. at $x/\bar{L} = 1$. This is not the case for structures that are distorted more severely. Thus a long period, $\bar{L}_{app}$, determined from the position of the first maximum in $\gamma_1(x)$ is only an apparent one, and it is always overestimated [130]. An overestimation of 20% ($\bar{L}_{app} \approx 1.2\,\bar{L}$) is not unusual.

### The First-Zero Method of Correlation Function Analysis.

For the purpose of a practical graphical evaluation of the linear crystallinity, Eq. (8.67) can be applied to a *renormalized* correlation function $\gamma_1(x/\bar{L}_{app})$. The method which has been proposed by Goderis et al. [162] is based on the implicit assumption that the first zero, $x_0$, of the real correlation function is shifted by the same factor as is the position of its first maximum, $\bar{L}_{app}$.

The idea is already described in the first paper of VONK and KORTLEVE ([159], p. 22) as a method to retrieve fit parameters. In their second paper ([160], p. 128) the authors state that inaccurate values are returned, if the found linear crystallinity is between 0.35 and 0.65.

The general inferiority of geometrical construction methods [162, 163] as compared to more involved methods which consider polydispersity has first been demonstrated by SANTA CRUZ et al. [130], and later in many model calculations by CRIST [165–167]. Nevertheless, in particular the first-zero method is frequently used. Thus, it appears important to assess its advantages as well as its limits. Validation can be carried out by graphical evaluation of model correlation functions [130, 165].

If the statistical model of a *paracrystalline stack* is assumed, it turns out that the renormalization attenuates the influence of polydispersity on the position of the first zero. In general, the first-zero method is more reliable than the valley-depth method, although it is not perfect. Even the first-zero method is overestimating the value of $v_l$. The deviation is smaller than 0.05, if the found crystallinity is smaller than 0.35. If bigger crystallinities are found, the significance of the determination is

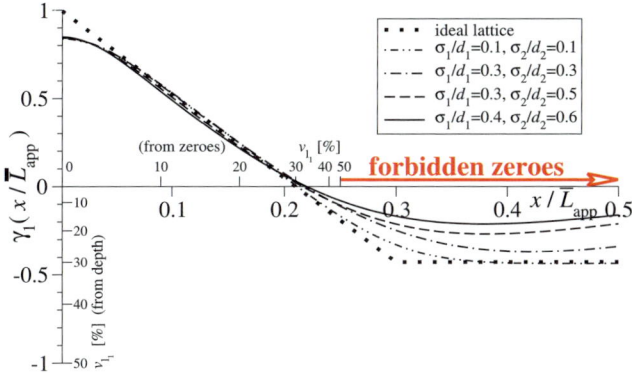

**Figure 8.22.** Testing the first-zero method for the determination of the linear crystallinity, $v_l$, from the linear correlation function, $\gamma_1 \left( x/\bar{L}_{app} \right)$ with $\bar{L}_{app}$ being the position of the first maximum in $\gamma_1 (x)$ (not shown here - but cf. Fig. 8.21). Model tested: Paracrystalline stacking statistics with Gaussian thickness distributions. The interval of forbidden zeroes is shown. An additional horizontal non-linear axis permits to determine the linear crystallinity directly. A corresponding vertical axis shows the variation of the classical "valley-depth method"

rapidly breaking down, and an individual demonstration of the error of determination becomes essential. In practice, insignificance can no longer be overlooked, if Eq. (8.69) applied to measured data does not return real solutions ("forbidden zeroes" in Fig. 8.22).

If the initial part of the correlation function exhibits significant deviations from a straight line, the proposers of the first-zero method recommend to carry out a linear regression (LRAT) [162] on the autocorrelation triangle. The problem of doing so is demonstrated in Fig. 8.21 and its discussion. Moreover, if the initial part of the correlation function does not only show a monotonous decay but discrete features, this is a strong indication of a topology that is not only polydisperse, but also heterogeneous[83]. In this case, a graphical correlation function analysis of isotropic data is of little significance anyway, and the study of uniaxially oriented material is recommended. Analysis may be performed by means of the CDF method (cf. Sect 8.5.5). If a low-noise scattering curve from isotropic material is at hand, it may be possible to separate components of a heterogeneous nanostructure by means of the IDF method (cf. Sect. 8.5.4) combined with model fits.

The first-zero method starts from the ideal lattice and Eq. (8.67). For the purpose of evaluation of scattering curves from polydisperse soft matter the ideal long period, $\bar{L}$, is replaced by $\bar{L}_{app}$, i.e. the validity of $\gamma_1 \left( v_l \left( 1 - v_l \right) \bar{L}_{app} \right) = 0$ is assumed. Because of the fact that the zero of a function is determined, not even a normalization of $\gamma_1 (x)$ is required [162]. Figure 8.22 displays the model data of Fig. 8.21 after the method-inherent renormalization $x \rightarrow x/\bar{L}_{app}$. Comparison with Fig. 8.21 shows that now

---

[83]No infinitely extended layers, several components with different topology (e.g. primary and secondary lamellar stacks)

the zeroes of the correlation functions with varying polydispersity are found close to the correct value $x_0/\bar{L}_{app} = 0.21 = 0.3\,(1 - 0.3)$. Vice versa, a good estimate for the linear crystallinity is obtained from the pair of roots which solve the quadratic relation

$$\frac{x_0}{\bar{L}_{app}} = v_{l_c}\,(1 - v_{l_c}). \tag{8.69}$$

If other statistical models of polydispersity should prove more appropriate than the paracrystalline stack, validations of the first-zero method may be carried out in analogy to the one presented here.

For anisotropic scattering patterns and the multidimensional case VONK ( [168] and [22], p. 302) has proposed to utilize a *multidimensional* correlation function. It is not frequently applied.

### 8.5.3 Isotropic Chord Length Distributions (CLD)

The isotropic chord length distribution (CLD) is of limited practical value if soft matter with only short-range order is studied. Nevertheless, the related notions have been fruitful for the development of new methods for topology visualization from SAXS data.

**Related Notions.**   Not only the 1D correlation function, but also the general 3D correlation function starts with a linear decay, and its series expansion

$$\gamma(r) = 1 - \frac{|r|}{\ell_p} + \dots. \tag{8.70}$$

was already given by POROD [18]. $\ell_p$ is the average chord length that has already been introduced on p. 112 in Eq. (8.23). Starting from this relation MÉRING and TCHOUBAR [118, 141, 169, 170] have derived that even the *distributions* of the individual segment lengths can be visualized by evaluation of an isotropic scattering pattern. They make use of the derivation theorem (p. 23, Eq. 2.39) applied to deliberate slicing directions of the structure and apply it twice. The two derivatives are distributed on each of the factors of the autocorrelation, $\Delta\rho^{*2}\,(r)$, and an ideal *edge enhancement* is accomplished. The result shows that the second radial derivative of the radial correlation function

$$\gamma''(r) = \frac{1}{\ell_p}\,(-2\,\delta\,(r) + g\,(r) + g\,(-r)) \tag{8.71}$$

is formed by two images of a *chord length distribution* (CLD), $g\,(r)$ and a $\delta$-distribution at the origin (Fig. 8.23). The CLD is made from an infinite series of segment distributions that starts with the homo-segment distributions, $\ell_1\,(r)$ and $\ell_2\,(r)$, for the domains of phase 1 and 2, respectively[84], followed by the di-segment distributions of the long periods, $-2\ell_{12}\,(r)$, and further out by the multi-segment distributions which describe the long-range arrangement of the particles in the material.

---

[84]Shape and size of the domains make these distributions.

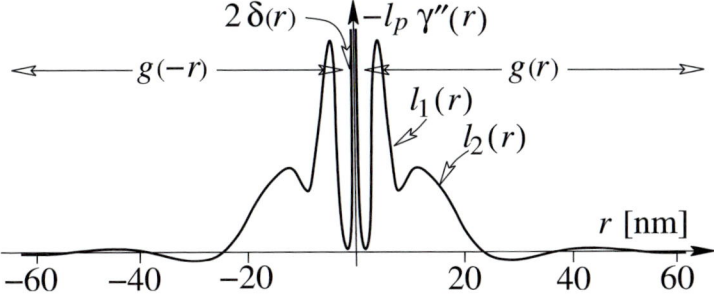

**Figure 8.23.** The chord length distributions $g(r)$ and $g(-r)$ found in the 2nd derivative $\gamma''(r)$ of the radial correlation function. The example shows $g(r)$ of a suspension of 10 wt.-% of silica (reproduced from a handout of DENISE TCHOUBAR)

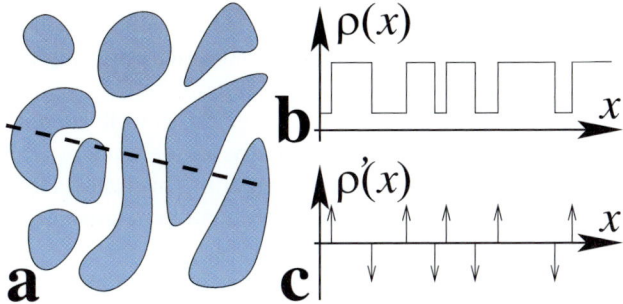

**Figure 8.24.** Demonstration of the edge-enhancement principle built into the chord length distribution. (a) Two-phase structure intersected by a straight line. (b) The density along the line. (c) The derivative of the density is a sequence of $\delta$-functions which are marking the positions of the domain edges

In the sketch taken from a handout of TCHOUBAR the distributions $\ell_1(r)$ and $\ell_2(r)$ are separated extraordinarily well.

The relation between structure and the chord distributions is readily established from considerations of topological density functions along a straight line traversing the material (Fig. 8.24). In Fig. 8.24a the respective sequence of chords is indicated. Figure 8.24b is a sketch of the corresponding density function, $\rho(x)$. Its first derivative, $\rho'(x)$ (Fig. 8.24c), is nothing but a sequence of $\delta$-functions put at the positions of the domain edges. Thus the edges are enhanced, and the autocorrelation $-\rho'(x) \star \rho'(-x) = g_p(x)$ is the partial CLD for the chosen special path through the topology.

For a general, isotropic and condensed multiphase material with short-range order, the CLD offers the best possible model-free visualization of the nanostructure. Nevertheless, the image does not show many details because of the inherent solid-angle average.

### 8.5.4 1D Interface Distribution Functions (IDF)

**Opportunities and Limits.** If we intend to obtain a clearer look on nanostructure than the one the CLD is able to offer, we can try to get rid of the orientation smearing – either by considering materials with a special topology (layer stacks), or by studying anisotropic materials.

If the scattering entities in our material are *stacks of layers with infinite lateral extension*, Eq. (8.47) is applicable. This means that we can continue to investigate isotropic materials, and nevertheless unwrap the 1D intensity of the layer stack. To this function RULAND applies the edge-enhancement principle of MÉRING and TCHOUBAR (cf. Sect. 8.5.3) and receives the interface distribution function (IDF), $g_1(x)$. Ruland discusses isotropic [66] and anisotropic [67] lamellar topologies.

For a layer-stack material like polyethylene or other semicrystalline polymers the IDF presents clear hints on the shape of the layer thickness distributions, the range of order, and the complexity of the stacking topology. Based on these findings inappropriate models for the arrangement of the layers can be excluded. Finally the remaining suitable models can be formulated and tested by trying to fit the experimental data.

As pointed out by STRIBECK [139, 171] $g_1(x)$ is, as well, suitable for the study of oriented microfibrillar structures and, generally, for the study of 1D slices in deliberately chosen directions of the correlation function. This follows from the Fourier-slice theorem and its impact on structure determination in anisotropic materials, as discussed in a fundamental paper by BONART [16].

In practical application to common isotropic polymer materials the IDF frequently exhibits very broad distributions of domain thicknesses. At the same time fits of the IDF curve to the well-known models for the arrangement of domains (cf. Sect. 8.7) are not satisfactory, indicating that the existing nanostructure is more complex. In this case one may either fit a more complex model[85] on the expense of significance, or one may switch to the study of anisotropic materials and display their nanostructure in a multidimensional representation, the multidimensional CDF. Complex domain topology is more clearly displayed in the CDF than in the IDF. The CDF method is presented in Sect. 8.5.5.

**Definition.** The interface distribution function

$$g_1(x) = -\left(\frac{d\rho_1(x)}{dx}\right)^{\star 2} = -k\,\gamma_1''(x) \tag{8.72}$$

is proportional to the 2nd derivative of the related 1D correlation function, $\gamma_1(r)$ (cf. Sect. 8.5.2).

**Computation.** $g_1(x)$ is computed from any 1D scattering intensity, e.g. $I_1(s_3)$

---

[85] A more complex model can be constructed from two components or a special sequence of (thick and thin) layers.

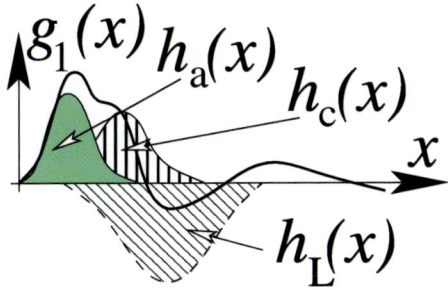

**Figure 8.25.** The features of a primitive interface distribution function, $g_1(x)$. The IDF is built from domain thickness distributions, $h_a(x)$ and $h_c(x)$, followed by the distribution of long periods, $h_L(x)$, and higher multi-thickness distributions

$$g_1(x) = -\mathscr{F}_1\left(4\pi s_3^2 I_1(s_3) - \lim_{s_3 \to \infty} 4\pi s_3^2 I_1(s_3)\right) \tag{8.73}$$

$$= -\mathscr{F}_1(G_1(s_3)) \tag{8.74}$$

by 1D Fourier transform. It is permitted to replace $I_1(s_3)$ by any[86] 1D projection $\{I\}_1(s_i)$ of a deliberate scattering pattern. The function which is subjected to the Fourier transform is identified as a 1D interference function, $G_1(s_3)$ (cf. page 140, Eq. 8.59).

**Interpretation.**    Similar to the CLD, $g(r)$ (constructed from a series of segment distributions), the IDF, $g_1(x)$ is a series of thickness distributions, $h_i(x)$. While the CLD lumps together all the segments that penetrate a domain in any deliberate direction, the IDF is more selective. Here a specific direction is chosen. Two examples: $x$ is the coordinate in the direction of the principal axis of the scattering entities; $r_3$ is the coordinate in fiber direction.

Thus, in the special case of a layer stack morphology, $g_1(x)$ is a series of thickness distributions (cf. Fig. 8.25). The series starts from the thickness distributions of "amorphous" and "crystalline" layers, $h_a(x)$ and $h_c(x)$, respectively. It is continued by the distributions of *aggregates* of adjacent layers, the first being an aggregate of one amorphous and one crystalline layer. The corresponding di-thickness distribution, $h_L(x) = h_{ac}(x) + h_{ca}(x) = 2h_{ac}(x)$ shows up with negative sign and represents the long periods. Thereafter we have the tri-thickness distributions $h_{aca}(x)$, $h_{cac}(x)$, and the following multi-thickness distributions.

Let us consider the other example. In an anisotropic material we select the fiber axis, $r_3$, project the intensity on this direction and compute an IDF. Then the meaning of the thickness distributions is quite similar as in the aforementioned example. Let us identify the first thickness distribution, $h_h(r_3)$, by a distribution of hard-domain thicknesses. Then the next thickness distribution, $h_s(r_3)$, is the thickness distribution

---

[86]That is, the direction of the projection may be chosen deliberately.

of the soft material in between, and the long period distribution is $h_L(r_3) = h_{hs}(r_3) + h_{sr}(r_3) = 2h_{hs}(r_3)$.

As we proceed from distribution to distribution within the series of thickness distributions, we observe that the functions are growing broader and broader. Moreover, their sign is alternating[87], and in a material with short-range order the IDF is already vanishing for relatively small values of $x$, $r_3$ or another chosen direction.

This observation is expected from theory, as the observed thickness distributions are exactly the functions by which one-dimensional short-range order is theoretically described in early literature models (ZERNIKE and PRINS [116]; J. J. HERMANS [128]). From the transformed experimental data we can determine, whether the *principal* thickness distributions are symmetrical or asymmetrical, whether they should be modeled by Gaussians, gamma distributions, truncated exponentials, or other analytical functions. Finally only a model that describes the arrangement of domains is missing – i.e., how the *higher* thickness distributions are computed from two principal thickness distributions (cf. Sect. 8.7). Experimental data are fitted by means of such models. Unsuitable models are sorted out by insufficient quality of the fit. Fit quality is assessed by means of the tools of nonlinear regression (Chap. 11).

> **Warning.** $g_1(0) \geq 0$ must hold. If in an experimentally determined curve $g(0)$ or $g_1(0)$ becomes strongly negative, there is a shortcoming in the pre-evaluation of the data. Probably the error originates from incorrect absorption correction or from errors in a manual evaluation of the Porod region. If a manual "deconvolution" of $g_1(x)$ is carried out, the areas of the peaks must conform to a zero-sum rule (cf. p. 158), and the centers of gravity of the peaks must conform to the obvious law of addition (e.g., $\bar{d}_c + \bar{d}_a = \bar{L}$ for the average crystalline thickness, the amorphous thickness and the long period). These constraints are not easily maintained manually, but can be programmed into a model function with little effort. Thus the constraints aggravate a manual evaluation of the IDF, but assist the deconvolution if methods of nonlinear regression are applied: even for rather diffuse IDFs unique deconvolutions can be found, if the type of the thickness distributions and the statistical model (Sect. 8.7) of domain arrangement is known. If distribution type and statistical model are varied, the results of the fits are discriminated by the quality of the match.

### 8.5.5 Anisotropic Chord Distribution Functions (CDF)

#### 8.5.5.1 Definition

The anisotropic multidimensional chord distribution (CDF) is an advancement of the IDF which is adapted to the study of highly anisotropic materials. CLD, IDF, and CDF are all based on the edge-enhancement principle devised by MÉRING and

---

[87] A negative long-period peak is always accompanied by two positive satellite peaks with each half the area (simplified zero-sum rule). Remember the alternating signs of the $\delta$-functions in Fig. 8.24 and have a look at p. 158.

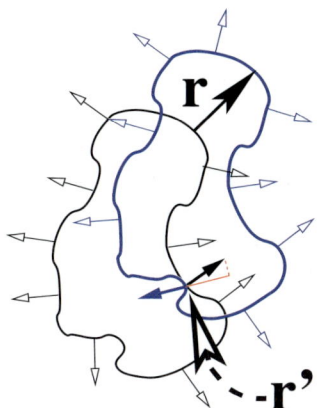

**Figure 8.26.** A particle–ghost autocorrelation of *gradient vectors* is generating the CDF. The vectors are emanating in normal direction from the surfaces of the particle and its ghost. The ghost is displaced by the vector **r**. The dashed arrow points at the position **r′**, at which a contribution to the CDF is generated. It originates from the scalar product of the two gradient vectors drawn in bold

TCHOUBAR. For the application to anisotropic scattering patterns STRIBECK [26] has extended this principle to a space of deliberate dimensionality. Available technology constricts its practical use to the scattering of materials with fiber symmetry, and the fiber-symmetrical CDF

$$z(r_{12}, r_3) = (\nabla\rho\,(r_{12}, r_3))^{\star 2} = k\,\Delta\gamma(r_{12}, r_3) \tag{8.75}$$

is closely related to VONK's multidimensional correlation function , $\gamma(r_{12}, r_3)$ ( [168] and [22], p. 302). One could think of synthesizing the CDF from a complete set of IDFs according to RULAND [66], but a viable algorithm for this path has not yet been found.

In space the 1D derivative $d/dx$ is replaced by the gradient $\nabla$, as is the second derivative $d^2/dx^2$ by the Laplacian $\Delta$ [26]. In analogy to the particle-ghost construction of the correlation function (cf. Figs. 2.4 and 8.24) the construction of the CDF can readily be demonstrated (Fig. 8.26). In a multiphase material the gradient field $\nabla\rho\,(\mathbf{r})$ is vanishing almost everywhere. Exceptions are the domain surfaces. They are densely populated with gradient vectors, the lengths of which are proportional to the heights of the density jumps.

### 8.5.5.2 Computation of the CDF for Materials with Fiber Symmetry

CDFs are computed from scattering data which are anisotropic and complete in reciprocal space. Thus the minimum requirement is a 2D SAXS pattern of a material with fiber symmetry taken in normal transmission geometry (cf. p. 37, Fig. 4.1). Required pre-evaluation of the image is described in Chap. 7.

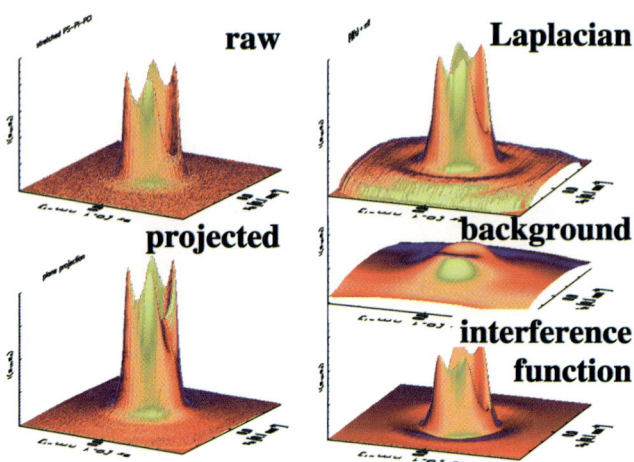

**Figure 8.27.** Steps preceding the computation of a CDF with fiber symmetry from recorded raw data: The image is projected on the fiber plane, the equivalent of the Laplacian in real space is applied, the background is determined by low-pass filtering. After background subtraction the interference function is received

**Transformation of the Pre-evaluated Image.** The image $I(s_{12}, s_3)$ is projected on the representative plane $(s_1, s_3)$ of the fiber scattering[88]

$$\{I\}_2(s_1, s_3) = 2 \int_0^\infty I\left(\sqrt{s_1^2 + s_2^2}, s_3\right) ds_2. \tag{8.76}$$

By means of this procedure our problem is not only reduced from three to two dimensions, but also is the statistical noise in the scattering data considerably reduced. Multiplication by $-4\pi s^2$ is equivalent to the 2D Laplacian[89] in physical space. It is applied for the purpose of edge enhancement. Thereafter the 2D background is eliminated by spatial frequency filtering, and an interference function $G(s_{12}, s_3)$ is finally received. The process is demonstrated in Fig. 8.27. 2D Fourier transform of the interference function

$$z(r_{12}, r_3) = -\mathscr{F}_2(G(s_{12}, s_3)) \tag{8.77}$$

finally yields the CDF, $z(r_{12}, r_3)$.

### 8.5.5.3 Relation Between a CDF and IDFs

Every radial, 1D slice through the center of a CDF

---

[88]The reason for this projection is that we are interested in the study of *slices* $\gamma(r_1, r_3) = \frac{1}{k}\mathscr{F}_2(\{I\}_2(s_1, s_3))$ *in real space*. So we must project in reciprocal space in order to reduce the fiber-symmetrical problem from three to two dimensions.

[89]Conceded – there is the alternative to apply a 3D Laplacian, but the corresponding procedure turns out not to be as stable as the 2D Laplacian when applied to experimental data.

$$\lceil z \rceil_1 \left( r_{\psi,\varphi} \right) = g_1 \left( r_{\psi,\varphi} \right)$$

is an IDF by definition. In the equation the slicing direction is indicated by a polar and an azimuthal angle, $\psi$ and $\varphi$, respectively. Of particular practical interest for the study of fibers is the cut of the CDF along the fiber axis,

$$\lceil z \rceil_1 \left( r_3 \right) = z(0, r_3) = g_1 \left( r_3 \right),$$

which describes the longitudinal structure of the material (cf. Sect. 8.4.3.2). In analogous manner the transversal structure (cf. Sect. 8.4.3.3) of the fiber is described by the slice

$$\lceil z \rceil_2 \left( r_{12} \right) = z(0, r_{12}) = g_2 \left( r_{12} \right)$$

of the CDF. A typical CDF of a highly oriented semicrystalline polymer material is shown in Fig. 8.28. Viewed from the top the domain peaks are visible, whereas viewing a CDF from the bottom shows the long periods peaking out.

### 8.5.5.4 How to Interpret a CDF

A CDF is interpreted in the same way as a CLD or an IDF. All these functions exhibit the probability distributions of domain size and arrangement. Clearer than a CLD is the IDF, because it does not contain an orientation average but exhibits the topology in a selected direction. Clearer than an IDF is the CDF, because it visualizes the nanodomain topology in space, i.e., in more than one direction.

**Uncorrelated Particles: Only Positive CDF Peaks.** Let us consider the simple example of identical, highly oriented cylinders which are randomly distributed in the material. Figure 8.29 demonstrates the scheme for the construction of the CDF assuming that the cylinder axis is parallel to the fiber axis. Two strong peaks on the meridian with almost triangular shape are characteristic for the cylinder. The *signal height* at a position **r** (i.e., at the position of the "glass rod" on the front peak in the sketch) is proportional to the area of contact between the cylinder and its displaced ghost. The basis length of the triangle is twice the *diameter* of the cylinder. The thickness of the triangle in meridional direction reflects the *polydispersity* of cylinder heights in the material. In addition, two weak diameter peaks are observed crossing the equator of the CDF. They are formed as the ghost is passing along the side of the particle.

The results of these considerations are readily extended from cylinders to lamellae: in the latter case the strong triangular peaks are wider, but closer to each other.

**Arrangement of Particles and the Corresponding Peaks.** If a CDF shows only positive peaks, the particles in the material are distributed at random[90]. There is no arrangement. Growing correlations are indicated by one or more triplets of peaks

---

[90]The reverse is only true for particles whose shape is *convex*, i.e., if the particles do not contain holes or indentations.

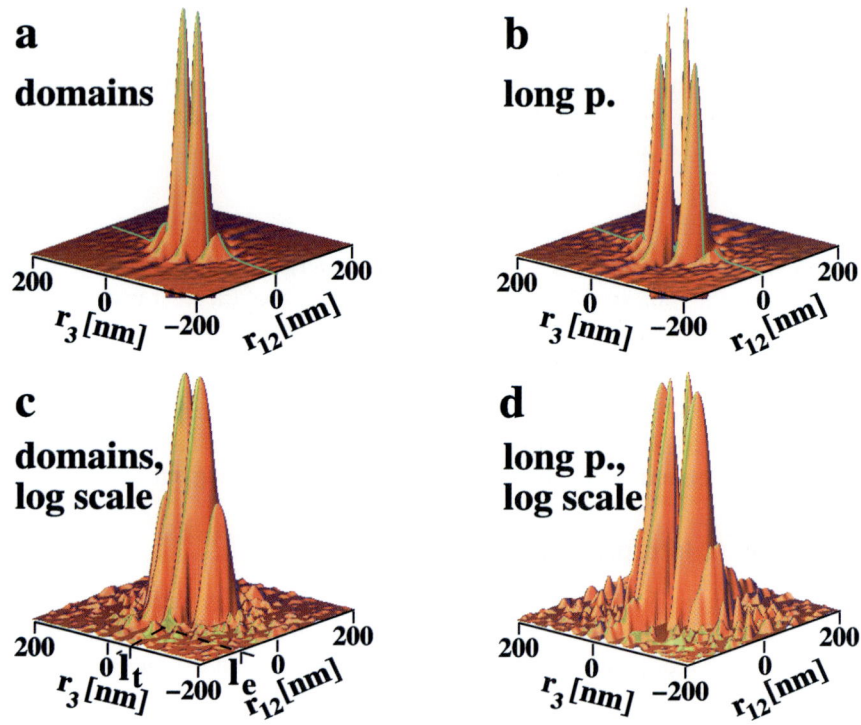

**Figure 8.28.** Demonstration of a CDF. Data recorded during non-isothermal oriented crystallization of polyethylene at 117°C. Surface plots show the same CDF: (a) Linear scale viewed from the top. (b) Linear scale viewed from the bottom. (c) Viewed from the top, logarithmic scale. Indicated are the determination of the most probable layer thickness, $l_t$, and of the maximum layer extension, $l_e$. (d) Viewed from the bottom, logarithmic scale. The IDF in fiber direction is indicated by a light line in (a) and (b) (Source: [56])

which do not change the integral of the CDF [172]. Let us demonstrate this general *zero-sum game of growing correlation* in one dimension by consideration of the IDF (Fig. 8.30). For every particle added to the structural entity, three additional peaks are observed. Their integral is zero[91].

From a practical point of view the sign of a peak in CLD, IDF, or CDF is described by the character of surface contact between particle and ghost[92]: if they contact each other in the normal way, the peak is positive; it is negative if they penetrate each other at the considered surface. Thus, positive peaks describe the size of par-

---

[91] I came across the zero-sum rule of correlation when I started to program models for structure fitting. Structure models which violate the zero-sum rule cannot be fitted to experimental data. Their convergence is poor.

[92] Up to now we have only discussed the correlation of a particle with *its own* ghost. In general, arrangement means that correlations between a particle and ghosts of *other particles from the same phase* are not extinguished by random annihilation.

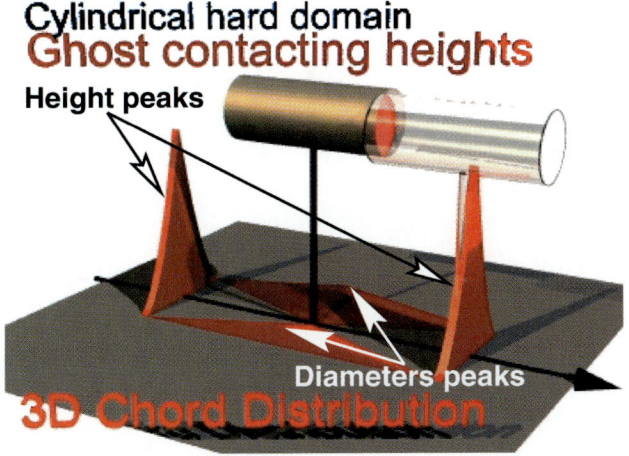

**Figure 8.29.** A particle-ghost displacement-principle governs the relation between structure and CDF. The height of the CDF signal is proportional to the *area of contact* between the particle and its ghost. A bold arrow in the base plane indicates the meridian (fiber direction)

ticles or super-particles. Negative peaks describe the space that is controlled by a particle or a super-particle from the structural entity. If the topology is addressed as a lattice, negative peaks show up at every repeat of the lattice constant (long period). If no long periods are detected, the structure describes an ensemble of uncorrelated particles. Every CDF analysis starts from such considerations.

### 8.5.5.5 Semi-quantitative CDF Analysis. An Example

**The Material of the Example.** Poly(ether ester) (PEE) materials are thermoplastic elastomers. Fibers made from this class of multiblock copolymers are commercially available as Sympatex®. Axle sleeves for automotive applications or gaskets are traded as Arnitel® or Hytrel®. Polyether blocks form the soft phase (matrix). The polyester forms the hard domains which provide physical cross-linking of the chains. This nanostructure is the reason for the rubbery nature of the material.

**Synopsis of Experiment and Results.** The material is irradiated during straining and relaxation. The example shows that a nanostructure which is hard to interpret from a series of scattering patterns may clearly reveal its complex domain structure after transformation to the CDF. Different structural entities are identified which respond each in a different way on mechanical load. The shape of the basic particles is identified (cylinders). The arrangement of the cylinders is determined. Thus the semi-quantitative analysis of the CDF provides the information necessary for the selection and definition of a suitable complex model which is required for a

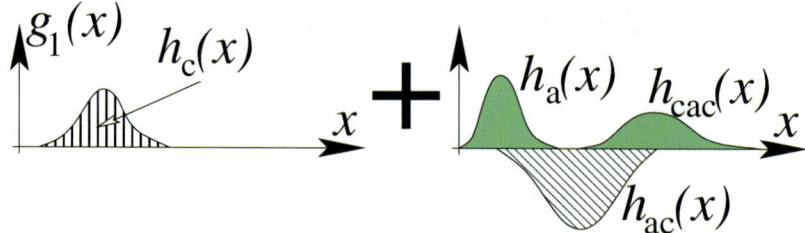

**Figure 8.30.** From particles to complex scattering entities in the IDF, the CDF or the CLD by growing correlation: An ensemble of uncorrelated particles exhibits only one homo-segment distribution (e.g., $h_c(x)$ representing crystallites). As next-neighbor correlations are growing, three segment distributions are added. *The integral of this triplet is zero.* Growing range of correlation adds further triplets

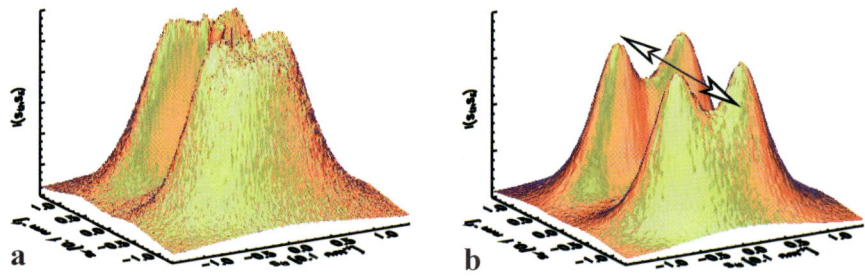

**Figure 8.31.** Fiber scattering of PEE 1000/43: (a) at an elongation $\varepsilon = 0.88$; (b) during relaxation from $\varepsilon = 0.88$. The fiber direction is indicated by a double-arrow. Visualized region: $-0.15\,\mathrm{nm}^{-1} \leq s_{12}, s_3 \leq 0.15\,\mathrm{nm}^{-1}$. $\varepsilon = (l - l_0)/l_0$, with $l_0$ and $l$ defined by the initial and the actual distance between two fiducial marks on the sample

complete quantitative analysis[93]. Even without a complete analysis mechanisms of structure evolution can be detected, if SAXS measurements are carried out *in situ* during processing by application of load (thermal, mechanical, . . . ).

Figure 8.31 shows central sections of two original SAXS patterns of PEE 1000/43[94] in strained and relaxed state. In the strained state (Fig. 8.31a) a "6-point-diagram" is detected. During relaxation (Fig. 8.31b) a well-separated "4-point-diagram" is observed. Interpretation of the patterns is restricted to description and speculation.

---

[93] In 3D a quantitative analysis still appears to be hopelessly laborious because of the complexity of the problem. On the other hand, a 1D quantitative analysis of only the longitudinal structure can be mastered (cf. Sect. 8.7).

[94] PEE's are commonly characterized by two numbers (e.g., 1500/50). The first number reports the minimum quantization of the polyether blocks (meaning "the polyether blocks are multiples of 1500 g/mol"), the second number indicates the mass fraction of the polyester hard phase (e.g., 50 wt.-% of polyester).

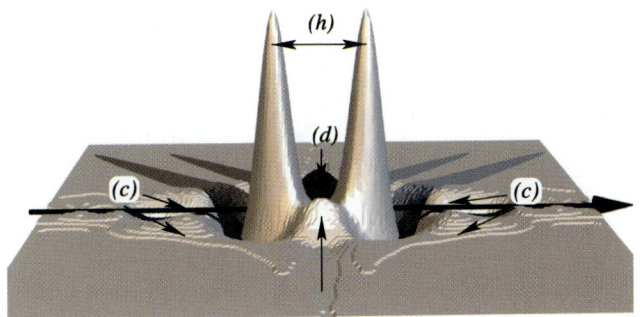

**Figure 8.32.** PEE 1000/43 at $\varepsilon = 0.88$. CDF $z(\mathbf{r})$. The domain peaks are pointing upwards: (h) cylinder-height peaks; (d) cylinder-diameter peaks; (c) inter-domain correlation peaks. Displayed region: $|r_{12}, r_3| \leq 40$ nm

In an original paper [173] the longitudinal structure has been studied quantitatively as a function of elongation. In a follow-up study [174] the 3D CDF has been computed and analyzed. Figure 8.32 shows the 3D CDF with fiber symmetry computed from the scattering pattern in Fig. 8.31. The straining direction $r_3$ is indicated by the long arrow in the basic plane. The observer is facing the domain peaks. Close to the origin the strong peaks on the meridian (h) mark the correlation between opposite faces of the basic domains. Two equatorial peaks (d) indicate the diameter of the domains. Because the height-to-diameter ratio is greater than 1, the basic domains can be approximated by *cylinders*. Four correlation peaks (c) are observed in an oblique angle with respect to the fiber axis. They indicate arrangement of domains. Their position shows that the closest neighbors of a cylinder are not found in straining direction, which would be indicative of a microfibrillar arrangement. Instead, the cylinders form a cluster with 3D short-range correlation. Such structural entities have been called a macrolattice by WILKE [175, 176]. The discussed peaks carry positive sign, because they describe chords that reach from the front face of a cylinder to the back face of a neighboring domain. The corresponding long periods show up as indentations observed at a shorter distance from the center, as they are measured "from front to front" of the domains. They are more easily observed after the CDF has been turned upside-down (Fig. 8.33). Obviously the long periods in fiber direction (a) are less pronounced than the long periods in oblique direction (b). Moreover, the CDF shows that the topology does not contain long-ranging correlations among domains. In fiber direction there is a long period of 25 nm (a), but already the size of the domain behind it can no longer be determined. On the other hand, the arrangement of domains in oblique direction (b) shows better correlation: here not only the long period, but also the size of the cylinder behind it can be determined (Fig. 8.32, (c)).

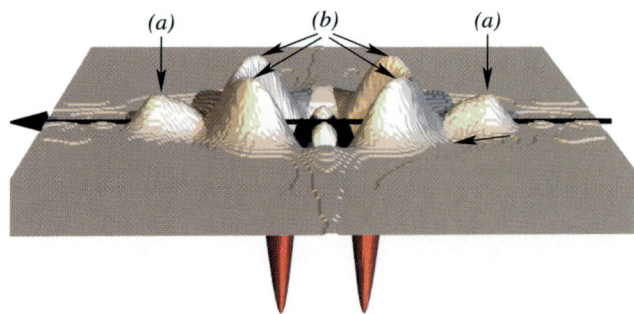

**Figure 8.33.** CDF $-z(\mathbf{r})$ of PEE 1000/43 at $\varepsilon = 0.88$. The long-period peaks are pointing upward: (a) long period to the next neighbor in straining direction; (b) stronger long period to the closest neighbor (in oblique direction)

During the beamtime another scattering pattern has been exposed after unloading the material. The respective CDF is shown in Fig. 8.34. Compared with the data from the strained state, the positions of the oblique long periods do not move (b). This finding indicates that the central cylinders are surrounded by domains which are rigidly coupled to them. In the scattering pattern such a structural entity is not easily discriminated from the 4-point diagram of a stack of inclined lamellae. In this respect the CDF is much clearer.

How should such rigid domain coupling work? In principle domains can only be rigidly coupled by a bridge of hard-phase material which has a different density. We know that the polyester hard-phase is semi-crystalline. So the observation is indicative for a structure in which the hard domains are subdivided into crystalline and amorphous zones. Thus a quantitative model of the structure would probably require to consider a third phase (*three-phase system*).

Finally we can compare the nanostructure in fiber direction after unloading with the nanostructure observed under mechanical load. The most striking variation is related to the strong long period (a), which is relaxing to half the value found in the elongated state. In addition to the strong long period, only in the unloaded material another long period is found (a′), for which even the 2nd order is visible. Thus the corresponding structural entities are built from domains with already a considerable range of correlation which are arranged along the straining direction. This is just the topological definition of a *microfibril* [157]. As the material becomes strained, the softer matter between the domains is elongated by different amounts and the longitudinal correlation gets lost. Thus the semi-quantitative analysis of the CDF returns a detailed view on the nanostructure evolution under load. More examples of

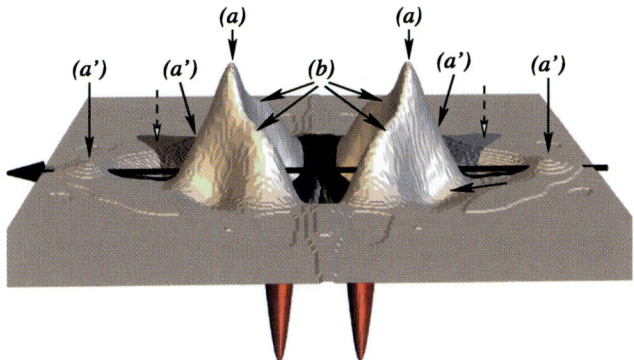

**Figure 8.34.** $-z(\mathbf{r})$ for PEE 1000/43 recorded during relaxation of the material from a first elongation to $\varepsilon = 0.88$. (a) strongest long-period in straining direction (13 nm) (dashed arrows with white head indicate the old positions of these peaks under strain); (a′): the best-correlated long-period in fiber direction (17 nm), because it shows a 2nd order; (b) oblique long period that is immovable in the straining experiment

the CDF method can be found in a growing number of original studies [56, 57, 177–186].

## 8.6 Biopolymers: Isotropic Scattering of Identical Uncorrelated Particles

**Overview.** Considerable research activities in the fields of isotropic SAXS and small-angle neutron scattering (SANS) are devoted to the investigation of ensembles of uncorrelated but identical or almost *identical complex particles*. Frequently these particles are studied in solution. Samples for such investigations must be supplied in a solution in which the particles do not aggregate.

The majority of the research is focused on colloidal and biological materials. In several textbooks [86, 101, 136, 187] the related methods are elaborated. Recent developments are considered in a review of SVERGUN and KOCH [188].

**Classical Analysis.** The classical analytical methods are even applicable for polydisperse samples and rest on the CLD (Sect. 8.5.3) and on VONK's [189] distance distribution function (DDF) ( [189–191]; [101] p. 168)

$$p(r) = r^2 \gamma(r).$$ (8.78)

Figure 8.35 shows for homogeneous identical spheres the radial correlation function (GUINIER and FOURNET [65] p. 12-19; LETCHER and SCHMIDT [192])

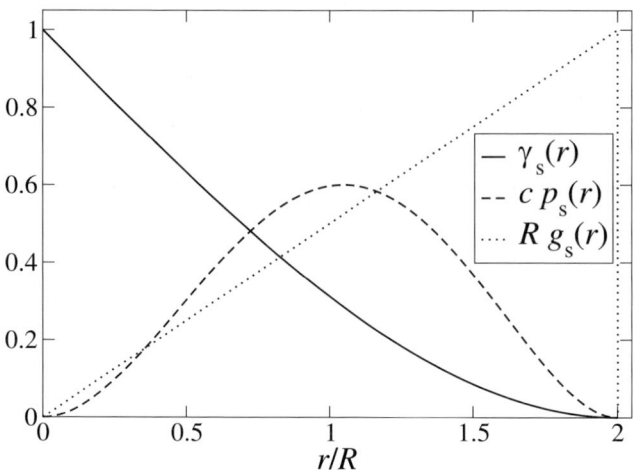

**Figure 8.35.** The homogeneous sphere of radius $R$. Radial correlation function, $\gamma_s(r)$, distance distribution function (DDF) $p_s(r)$ and chord length distribution (CLD) $g_s(r)$

$$\gamma_s(r) = 1 - \frac{3}{4}\left(\frac{r}{R}\right) + \frac{1}{16}\left(\frac{r}{R}\right)^3 , \qquad (8.79)$$

the DDF, $p_s(r)$, and the CLD [20] $g_s(r) = -\ell_p\,\gamma_s''(r)$ . All these functions are restricted to the interval $0 < r/R \leq 2$. Their support is finite. We observe that in $p_s(r)$ the maximum extension of the particle is visualized much clearer than in $\gamma_s(r)$. $p_s(r)$ is positive everywhere, because the sphere is homogeneous and convex.

Finite support of $\gamma(r)$, $p(r)$, and $g(r)$ is characteristic for a structure made from uncorrelated particles. Thus the related scattering curve is bandlimited (cf. p. 25). For analysis this property is advantageously utilized.

$g_s(r)$ exhibits a *discontinuity* at $r = 2R$. Such discontinuities must be expected both for monodisperse structure of identical particles or for almost perfectly arranged structure. As a consequence, the reconstruction of $g_s(r)$ from experimental data requires extreme accuracy both during measurement and during data pre-evaluation (BURGER and RULAND [20]). The discontinuity problem diminishes the practical value of structure visualization methods (Sect. 8.5) in the fields of biological materials and of highly ordered nanostructures (Sect. 8.8).

### Ab-Initio Methods for Shape Reconstruction of Identical Particles.

On the next level of analysis the generality is restricted to problems from the field of life sciences. In this case the dissolved particles (enzymes, proteins) are identical, in general. It is the aim to reconstruct the shape of the single particle from the isotropic SAXS curve. Uniqueness of a found solution cannot be guaranteed – in particular if a compact particle shape is returned, the ambiguity is rather high. If the reconstruction yields elongated or open particles, the significance is higher. The value of the method is highest if the result of the analysis shows that a previously proposed (protein) structure *does not match* the observed SAXS curve.

For any method of structure reconstruction by ab-initio methods additional assumptions must be made. The *multipole expansion*[95] method of HARRISON, STUHR-MANN, and SVERGUN ([86], Sect. 5.3; [101], Chap. 6) assumes homogeneous internal density. The shape of the scattering curve is fitted by varying the envelope of the particle.

More powerful are the finite-element methods. Here the shape sought-after is approximated by an agglomerate of many small elements. Different elementary bodies are utilized. Beads (i.e., small spheres) are easily handled, because their form factor does not change upon solid-angle average [193] and the DDF is readily established (GLATTER in [101], p. 160, SVERGUN [194]). A modern finite-element method has been developed for the study of proteins (SVERGUN [194, 195]): the "Dummy Residues Method" represents each amino acid residue by a spherical bead of homogeneous and identical density. The selectivity of the method is founded on built-in knowledge concerning the chemical structure of all proteins: the distance between two amino acid residues is 0.38 nm.

A programming package PRIMUS for the evaluation of isotropic SAXS patterns is offered by SVERGUN [196]. Although the focus is on biopolymers, it can also evaluate general particle scattering.

**Power and Limits of the SAXS Methods.** This field of SAXS is in competition with the field of protein crystallography. The spatial resolution of the SAXS method is limited ($> 0.5$ nm), whereas structures determined by protein crystallography are exact up to fractions of Ångstrøms. On the other hand, the protein crystallography is unable to study "living" proteins under almost physiological conditions. Moreover, kinetic processes can be monitored by SAXS but cannot be studied by means of protein crystallography.

If mixtures and assemblies of different kinds of particles are studied, a *shape reconstruction* is no longer possible from a study of isotropic SAXS in solution [188].

## 8.7 Quantitative Analysis of Multiphase Topology from SAXS Data

The quantitative analysis of a multiphase topology comprises the formulation of structure models and the fitting of measured data. Fitting is discussed in Chap. 11. In this section the setup of topological models is discussed. The problem arises from the fact that most structural models of particle correlation are anisotropic and the visualization of structure in anisotropic materials by means of the CDF shows that suitable models must be rather complex. Thus a direct fit of anisotropic data would require fitting of a measured 3D or 2D function by a complex model. Both the effort to setup such models, and the computational effort to fit the data are very high.

---

[95]The multipole expansion is defined on p. 194 in Eq. (9.3).

As a way out it could be considered to record and to fit isotropic scattering data, but for strongly polydisperse materials this is not a good solution, because then a complex structural model must be fitted to scattering data that are, additionally, blurred by solid angle averaging yielding low-significance results. Nevertheless, there is a promising strategy, if anisotropic data are available. Instead of a multidimensional analysis, an interesting aspect of the topology can be extracted by means of 1D projection (cf. Sect. 8.4.3.2), and the corresponding 1D problem can be fitted more easily. Even a 2D projection may be suitable (example in Sect. 8.7.1.2). For this purpose one- and two-dimensional models with increasing complexity must be constructed.

### 8.7.1 Models for Uncorrelated Polydisperse Particles

### 8.7.1.1 Polydisperse Layers and 1D Particles

Let us consider the last stage of melting of a semicrystalline polymer. It is frequently postulated that in this stage only few single, uncorrelated crystalline lamellae should be found. In principle this postulate can directly be tested by inspection of the IDF (for isotropic materials) or of the CDF (for highly oriented materials). If such a transformation shall be avoided, one may fit a model function directly to the scattering intensity, which is taken for the particle scattering of an ensemble of lamellae. Although not frequently done, one should consider polydispersity and not assume that all the layers show the same thickness.

For an ensemble of uncorrelated 1D particles (cylinders, layers) with a Gaussian[96] particle thickness distribution the 1D scattering intensity is [197]

$$I_1(s_3) = \frac{A_{P_1}}{s_3^2} \left(1 - \cos\left(2\pi \bar{d}_c s_3\right) \exp\left(-2\pi^2 \sigma_c^2 s_3^2\right)\right). \tag{8.80}$$

Here $\bar{d}_c$ is the average thickness and $\sigma_c^2$ is the variance of the particle thickness distribution modeled by a Gaussian. $A_{P_1}$ is the 1D Porod asymptote (cf. p. 125, Table 8.3). The particle thickness distribution considers polydispersity (cf. Chap. 1).

If the structural entities are lamellae, Eq. (8.80) describes an ensemble of perfectly oriented but uncorrelated layers. Inversion of the LORENTZ correction yields the scattering curve of the isotropic material $I(s) = I_1(s) / (2\pi s^2)$. On the other hand, a scattering pattern of highly oriented lamellae or cylinders is readily converted into the 1D scattering intensity $\{I\}_1(s_3)$ by 1D projection onto the fiber direction (p. 136, Eq. (8.56)). The model for the 1D intensity, Eq. (8.80), has three parameters: $A_{P_1}$, $\bar{d}_c$, and $\sigma_c$. For the nonlinear regression it is important to transform to a parameter set with little parameter-parameter correlation: $A_{P_1}$, $\bar{d}_c$, and $\sigma_c / \bar{d}_c$. When applied to raw scattering data, additionally the deviation of the real from the ideal two-phase system must be considered in an extended model function (cf. p. 124).

With respect to the IDF

---

[96]For a generalization cf. p. 180, Eq. (8.105)

$$g_1(r_3) = \frac{A_{p_1}}{2\sqrt{2\pi\sigma_c^2}}\left(\exp\left[-\frac{1}{2}\left(\frac{r_3-\bar{d}_c}{\sigma_c}\right)^2\right]+\exp\left[-\frac{1}{2}\left(\frac{r_3+\bar{d}_c}{\sigma_c}\right)^2\right]\right) \quad (8.81)$$

the model function is simply the Gaussian particle thickness distribution showing up both on the positive and on the negative branch of the $r_3$-axis.

We have chosen Gaussian thickness distributions, because structure visualization by means of IDF or CDF exhibits thickness distributions that frequently look very similar to Gaussians[97]. The presented relations for the 1D intensity and the IDF are the basic relations for many 1D structure models, comprising the general analysis of materials made from layers, highly oriented microfibrillar materials, and the direction-dependent analysis of anisotropic materials.

### 8.7.1.2 Uncorrelated Particles in 2D: Fibril Diameters in Fibers

**Experiment and Problem.** An equatorial streak in a highly oriented fiber pattern indicates rod-shaped structural entities oriented with their axes parallel to the fiber direction. Such entities may be microfibrils, added nucleating agents[98], needle-shaped voids [198, 199], or needle-shaped crystals ("shish"). Figure 8.36 shows the pseudo color representation of such a scattering image. It has been recorded during straining of a poly(ether ester). A blind central spot has been filled by 2D extrapolation. There is little discrete scattering along the ridge of the streak.

### Complications and Their Solution

**Imperfect Orientation.** If the streak were fanned out (cf. p. 202, Fig. 9.6), the orientation smearing must first be extinguished (Sect. 9.7) before the scattering of the perfectly oriented structural entities is retrieved.

**Distribution of Rod Lengths.** If the distribution of rod lengths shall be studied, the smearing of the equatorial streak by the primary beam profile must be eliminated[99]. After that the 1D scattering intensity is computed by means of Eq. (8.56) and fitted to the respective 1D model (e.g., Eq. (8.80)) from Sect. 8.7.1.1. Be careful. The rods may, in fact, not be stretched out perfectly but only resemble long "worms" instead. In this case the determined rod length is not the true length but only

---

[97]Sometimes the distributions look asymmetrical. In this case it is possible to switch from *Gaussians* to the *Mellin convolution of two Gaussians* [125]. The Gaussian has the shortcoming that broad Gaussians ($\sigma_c/\bar{d}_c > 0.3$, cf. p. 20) predict negative domain thicknesses. For practical application this has no consequences, because the autocorrelation principle reflected in the central-symmetry of Eq. (8.81) implicates that only the modulus of the domain thickness is accounted for. Nevertheless, the overflow of the distribution tail changes the effective shape of the function. To avoid this shortcoming, one may, for example, switch from Gaussians to *gamma distributions* (cf. p. 180).

[98]Many industrial semi-crystalline polymer materials like polypropylene, polyamides, or polyesters contain nucleating agents or clarifiers which form needle-shaped aggregates already in the polymer melt.

[99]For this purpose the pattern is desmeared using the measured primary beam. For a less involved treatment it may be sufficient to know the integral width of the primary beam profile in fiber direction.

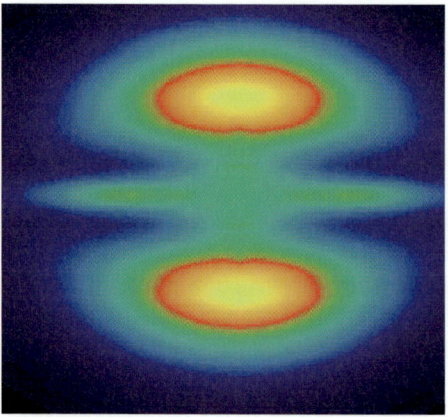

**Figure 8.36.** SAXS pattern of a thermoplastic elastomer during straining. The thin horizontal line in the center is called an equatorial streak. In this case it is well-separated from the long-period peaks above and below

the *persistence length* of the worms, and the polydispersity parameter describes the width of the persistence-length distribution.

**2D Streak Analysis.** The rod scattering from the equatorial streak, $I_R(\mathbf{s})$, is extracted from the scattering pattern [200] and projected

$$\{I_R\}_2(s_{12}) = \int_{-\infty}^{\infty} I_R(\mathbf{s})\, ds_3 \tag{8.82}$$

on the cross-sectional plane of the fiber. As already discussed (p. 139), this scattering curve contains the information on the transversal structure of the fiber as sketched in Fig. 8.18. The sketch shows the structure in 3D. We have *projected* the scattering data, thus the resulting curve describes – in the averaged fiber cross-*section* – the size distribution of the rod cross-sections and their mutual correlation. For low correlation such a 2D structure has been named a hard-disc fluid (COHEN and THOMAS [201]).

**Is this an Uncorrelated Hard-Disc Fluid?** In order to answer this question we compute a 2D CLD and test, whether the function is positive everywhere. In this case the equatorial streak can be considered pure particle scattering.

The CLD is computed from an interference function (cf. p. 139). In order to consider the non-ideal two-phase system we either carry out a classical Porod analysis according to Table 8.3 for the case of a 2D projection in order to retrieve the interference function

$$G_2(s_{12}) = (\{I_R\}_2(s_{12}) - I_{Fl})\, s_{12}^3 / \exp\left(-4\pi^2 \sigma_z^2 s_{12}^2\right) - A_{P_2}, \tag{8.83}$$

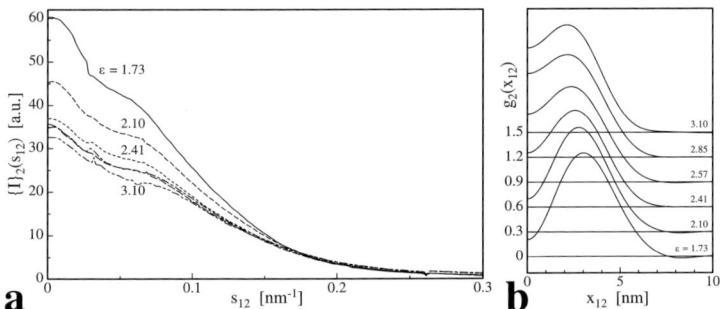

**Figure 8.37.** Equatorial streak analysis of a poly(ether ester) during straining as a function of elongation, $\varepsilon$. (a) Intensity projected on the cross-sectional plane (b) 2D CLD computed from the projected intensity

or we extract the interference function automatically by means of spatial frequency filtering (p. 140). The sought-after CLD of the fiber cross-sections

$$g_2(r_{12}) = \pi \int_0^\infty \left( J_0(2\pi r_{12}s_{12}) - J_2(2\pi r_{12}s_{12}) \right) G_2(s_{12}) \, ds_{12} \qquad (8.84)$$

then is the 2D Fourier transform of $G_2(s_{12})$ [200]. $J_i()$ is the Bessel function of the first kind and order $i$ ( [4], p. 358).

Figure 8.37a shows the projected intensity[100] as a function of the elongation of the material. The resulting 2D CLDs are presented in Fig. 8.37b. Equatorial streaks are observed from an elongation of $\varepsilon = 1.73$. Only at this elongation there are some faint correlations among the microfibrils: the CLD becomes slightly negative at $r \approx$ 8 nm. For higher elongation the structure in the fiber cross-section conforms to a hard-disc fluid.

### Model Fit or Direct Evaluation?

**Model Fit or Direct Evaluation?**  Similar to the 1D case we can now fit a model to the projected intensity or to the CLD. What we get in this case is the needle-diameter distribution of the microfibrils. Nevertheless, there are two other possibilities to directly evaluate the data. We consider polydispersity by allowing for varying hard-disc diameter. If we assume that the shape of each disc is circular, the CLD of an uncorrelated hard-disc fluid is the Mellin convolution of the intrinsic chord distribution, $g_c(r_{12})$, of an "ideal disc of diameter 1" and the diameter distribution, $h_D(D)$ which characterizes the structure. The definition of the Mellin convolution (TITCHMARSH [202], S. 53; MARICHEV [203]; [86], S. 304) is

---

[100]The discontinuities at $s_{12} \approx 0.03\,\text{nm}^{-1}$ indicate the match between measured and extrapolated intensity. Because in the next step the intensity is multiplied by $s_{12}^3$, it is not necessary to spend more effort on a smooth continuation.

$$g(r) = \int_0^\infty h_D(x)\, g_c\left(\frac{r}{x}\right)\frac{dx}{x} \tag{8.85}$$

$$:= g_c(r) \odot h_D(r). \tag{8.86}$$

$g_c(r_{12})$ is analytical. So we can avoid to fit models, but "deconvolute" the Mellin convolution numerically and obtain $h_D(D)$. Even better: because of a very simple relation among the moments of a Mellin convolution [125]

$$\mu_i'(g \odot h) = \mu_i'(g)\,\mu_i'(h) \tag{8.87}$$

we can directly determine the interesting structural parameters (average needle[101] diameter, width of the needle-diameter distribution) by numerical computation (i.e., integration) of moments followed by simple moment arithmetics [200].

The equation for the intrinsic chord distribution of a circular disc of diameter 1 has first been published by P. W. SCHMIDT [204]

$$g_c(r_{12}) = \frac{r_{12}}{\sqrt{1 - r_{12}^2}}\, Y_H(1 - r_{12}). \tag{8.88}$$

Here

$$Y_H(x) = \begin{cases} 1 & /x \le 0 \\ 0 & /x > 0 \end{cases} \tag{8.89}$$

is the Heaviside function. Thus the problem is reduced to an invertible integral transform. The first method of structure determination is based on numerical Mellin deconvolution of $g_2(r_{12})$ yielding $h_D(D)$. After the deconvolution, structural parameters have to be determined by numerical computation of moments [200].

The second method is more elegant, because it only involves the numerical computation of moments (cf. Sect. 1.3) of the smeared CLD $g_2(r_{12})$ followed by moment arithmetics [200]. The first step is the computation of the Mellin transform[102] of the analytical function $g_c(r_{12})$ which we have selected to describe the needle diameter shape. This is readily accomplished by Mathematica® [205]. Because the Mellin transform is just a generalized moment expansion, we retrieve for the moments of the normalized chord distribution of the unit-disc[103]

$$\mu_i'(g_c) = \frac{1}{\sqrt{\pi}}\frac{\Gamma((i+2)/2)}{\Gamma((i+3)/2)},$$

i.e., explicitely for the first 5 moments

---

[101] The diameter of the (3D) needle is the (2D) hard disc.

[102] The Mellin transform is defined by $\mathcal{M}(g(x))(s) = \int_0^\infty x^s g(x)\,(dx/x)$. Substituting the variables in the Mellin transform by their logarithms, the Fourier transform is obtained [202].

[103] Conceded – this "unit disc" is not the unit disc of mathematics (with *radius* 1), but of materials science (with *diameter* 1).

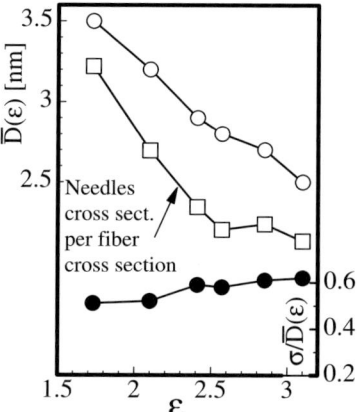

**Figure 8.38.** Structural parameters of an ensemble of needle-shaped soft domains in a poly(ether ester) as a function of elongation $\varepsilon$. $\bar{D}$ (open circles) is the average needle diameter, $\sigma/\bar{D}$ (filled circles) is the relative standard deviation of the needle-diameter distribution. Square symbols demonstrate the lateral compressibility of the soft needles during elongation

| $i$ | 0 | 1 | 2 | 3 | 4 |
|---|---|---|---|---|---|
| $\mu_i'(g_c)$ | 1 | $\pi/4$ | $2/3$ | $(3\pi)/16$ | $8/15$ |

Now using this table and Eq. (8.87) the moments of $h_D(D)$ are directly computed from the moments of $g_2(r_{12})$. Structural parameters directly determined by this method are presented in Fig. 8.38. Obviously, the average diameter of the soft domains is decreasing first linearly with increasing elongation, although for rubber elastic materials one would expect a decrease only according to $\bar{D}(\varepsilon) = \bar{D}_0 / \sqrt{\varepsilon + 1}$. This "observed deviation" from rubber elasticity is not a real finding, but only the consequence of oversimplification, because up to now we have neither considered the polydispersity of the needle cross-sections nor investigated changing shape of the needle-diameter distribution. Figure 8.37 shows that during initial elongation the fraction of thin needles is increasing strongly, as the initial compressibility of the soft needles is high (Fig. 8.38, squares). Only beyond $\varepsilon > 2.5$ the relative cross-sections of soft needles per fiber diameter stays constant. This transition can be explained by the well-known strain-induced crystallization of the polyether.

**Consideration of Weak Correlation.**    If an inter-needle correlation can no longer be disregarded, two published methods for the analysis of weakly correlated particles may be considered: (COHEN and THOMAS [201]; POROD [206]).

### 8.7.1.3 Uncorrelated Polydisperse Homogeneous Spheres

Let us now consider the 3D equivalent of the aforementioned example: an ensemble of uncorrelated homogeneous spheres – with polydispersity, meaning that the observed CLD

$$g_{obs}(r) = g_{s1}(r) \odot h_D(r)$$

is obtained by Mellin convolution from the CLD of the unit-sphere

$$g_{s1}(r) = 2rY_H(1-r) \tag{8.90}$$

and a sphere-diameter distribution $h_D(r)$. Again of physical interest are the moments of $h_D(r)$ that can readily be combined to retrieve physical parameters like the average sphere diameter, the breadth, or the skewness of $h_D(r)$. From the Mellin transformation

$$\mathcal{M}(g_{s1}(r))(s) = 2/(1+s)$$

of the unit sphere we obtain for its moments[104]

$$\mu_i'(g_{s1}) = 2/(2+i). \tag{8.91}$$

**Check.** Do the moments reflect the requirement that $g_{s1}$ is normalized to 1 and that the average chord length of a sphere with unit diameter is $2/3$?

Bearing in mind the aim to determine structural parameters, we resort to the measured CLD, $g_{obs}(r)$. By means of numerical integration we compute some of its moments, $\mu_i'(g_{obs})$. For the sought-after moments of the sphere-diameter distribution, $h_D(r)$, we again have according to Eq. (8.87)

$$\mu_i'(h_D) = \mu_i'(g_{obs})/\mu_i'(g_{s1}),$$

and the $\mu_i'(g_{s1})$ are analytical (Eq. (8.91)). The most important structural parameter is the number *average*[105] *sphere diameter*

$$\bar{D} = \mu_1'(h_D)/\mu_0'(h_D).$$

The common measure for the *breadth of the sphere-diameter distribution* is the standard deviation (cf. p. 5, Eq. (1.8)).

### 8.7.1.4 Inhomogeneous Spherical Particles

A concise approach for the analysis of isotropic scattering curves of spherical and cylindrical *particles with a radial density profile* has been developed by BURGER [207]. In practice it is useful for the study of latices and vesicles in solution.

---

[104]Moments are obtained from a Mellin transform by changing from $s$ to $i+1$.

[105]As $g_{obs}(r)$ is normalized to 1 by definition, here the division by $\mu_0'(h_D)$ is unnecessary – but it demonstrates the principle and it is safe.

### 8.7.2 Stochastically Condensed Structure

**Introduction.**   After we have discussed examples of uncorrelated but polydisperse particle systems we now turn to materials in which there is more structure – discrete scattering indicates correlation among the domains. In order to establish such correlation, various structure evolution mechanisms are possible. They range from a stochastic volume-filling mechanism over spinodal decomposition, nucleation-and-growth mechanisms to more complex interplays that may become palpable as experimental and evaluation technique is advancing.

In the borderland between diffuse and discrete scattering there are fundamental questions like: is it possible to observe a long period in a SAXS pattern without the structure of a lattice (i.e., some "repeat unit")? The first indications for a positive answer have been published in 1974 by KILIAN and WENIG [208] who observed long period peaks in polyethylene material, in which correlation was restricted to only two layers. Concerning the underlying structure evolution mechanism, SCHULTZ et al. [209] have inferred a predominantly stochastic *crystallization* mechanism from time-resolved SAXS studies. Concerning *melting* of polyethylene, a study of SCHULTZ et al. [210] has indicated a completely random melt-out of crystallites based on SAXS data from isotropic polyethylene. In the view of a mathematical concept introduced in this section, such a melting mechanism is a "random de-parking process".

**Random Population and Order.**   In Sect. 8.7.1 we have discussed pure particle scattering. Such scattering is observed if the sample volume is sparsely populated with domains and an order establishing mechanism is missing. Nevertheless, above a certain population density of particles, newcomers must arrange with their neighbors. They can only settle *where* there is enough space left. If we find that such "arrangement" is sufficient for a long period peak to show up in the scattering, the next question is: how is it possible to determine if observed formation of discrete scattering is governed by an ordering process or if it is only a side effect of crowding? In practice such problems arise if we study the isothermal crystallization of a polymer, during which the volume is continuously populated with lamellae.

**Scattering and Disorder.**   For structure close to random disorder the SAXS frequently exhibits a broad shoulder that is alternatively called "liquid scattering" ([206]; [86], p. 50) or "long-period peak". Let us consider disordered, concentrated systems. A poor theory like the one of POROD [18] is not consistent with respect to disorder, as it divides the volume into equal lots before starting to model the process. He concludes that statistical population (of the lots) does not lead to correlation. Better is the theory of HOSEMANN [158,211]. His distorted structure does not pre-define any lots, and consequently it is able to describe (discrete) liquid scattering. The problems of liquid scattering have been studied since the early days of statistical physics. To-date several approximations and some analytical solutions are known. Most frequently applied [201, 212–216] is the PERCUS-YEVICK [217] approximation of the ORNSTEIN-ZERNIKE integral equation. The approximation offers a simple descrip-

**Figure 8.39.** Structure formation by random crowding is the issue of the "car parking problem" of mathematics

tion of the scattering from a topology of poorly ordered particles. On the other hand, this model is such a coarse approximation that it cannot easily be interpreted in terms of structure evolution mechanisms [218].

**Order by Crowding *vs.* Ordering Mechanism – A Summary.**   As a result of the following considerations, the correlations introduced by a crowding mechanism can be discriminated from the correlations grown from a genuine ordering mechanism: the stochastic mechanism can only generate next-neighbor correlations. Thus *a second minimum* visualized in the CLD, IDF, or CDF proves an ordering mechanism. Vice versa – is the finding of only next-neighbor correlations sufficient to prove random crowding? The answer is no. Nevertheless, the *shape* of the peaks in CLD, CDF, or IDF may indicate a pure random crowding process. It can be proven that after a purely stochastic crowding the thickness distributions are not Gaussians, but (strongly asymmetrical) truncated exponentials. We thus can tell apart, whether the last "parking" crystallites (cf. Fig. 8.39) keep equal distance to their left and right neighbor[106], or whether they do not care at all where they park in the gap [184]. If the polydispersity (i.e., the width variation of the cars) is low, theory [219] returns an ultimate "crystallinity" of 75% for the stochastic crystallization, the so-called Rényi limit. Even for the correlation function of the stochastic structure an analytical series has been found [218]. Unfortunately it is poorly converging and infinite.

**The Car Parking Problem.**   The fundamental 1D stochastic process applied to crystallization is described in simple words using the 1D shish-kebab model: crystalline layers (kebab) of almost identical thickness are formed at random positions along a backbone (shish). The process continues, until the last gap of sufficient width has been filled. The jamming limit is reached. For layers of equal thickness this process is known by the name "car parking problem". The corresponding kinetics and some of the properties of the resulting structure have first been studied by RÉNYI [219, 220]. For example, at the *jamming limit* the theoretical ultimate volume crystallinity of stochastic crystallization

$$v_{cs}^{\infty} = 2 \int_0^{\infty} \exp\left[-t - 2 \int_0^t \frac{1 - \exp(-u)}{u}\, du\right] dt = 0.747\,597\,92 \qquad (8.92)$$

is reached. It is called the Rényi limit. It says that no more cars can be parked, as soon as 75% of the length of a parking lane have been occupied without consid-

---

[106]This would indicate a late short-range ordering mechanism

eration. Related problems are subsumed under the term *random sequential adsorption* (RSA) (review by EVANS [221]). The "structure" originating from random car parking shows a long-period maximum. Even its structure factor and its correlation function can be described analytically [218].

**Stochastic Packing and the Truncated Distribution of Amorphous Thicknesses.** At the *beginning* of stochastic crystallization the distances $h_a(r_3) = c$ between the crystal lamellae are uniformly distributed. As the process is proceeding the number of positions is decreasing, at which a new crystal can be formed. Ultimately all gaps that are sufficiently wide have been occupied. The distribution $h_a(r_3)$ of the amorphous layers has changed its shape and has turned into a truncated probability distribution [222, 223]: its support is limited to a narrow interval[107]. Analytical deductions of the shape of $h_a(r_3)$ are unknown. Nevertheless, in various fields of science heuristic descriptions are found [222–225], which generally report that these distributions can be described by simple exponential functions. Computer simulations of the iterated stochastic crystallization yield the same result [184, 226]. By fitting of exponentials to the data of a series of simulations one obtains the result [184]

$$h_a(u) = \frac{1}{N} \exp\left(-\frac{2}{3}\pi\sqrt{u}\right), \quad u \in (0,1), \tag{8.93}$$

with $u = r_3/\bar{d}_c$ and the crystallite thickness $\bar{d}_c$. After normalizing $\int h_a(u)\,du = 1$ the total probability to 1 it follows

$$N = \frac{9\left(\exp(2\pi/3) - 1\right) - 6\pi}{2\pi^2 \exp(2\pi/3)}, \tag{8.94}$$

and for the center of gravity of the distribution of amorphous thicknesses we have

$$\langle h_a(u) \rangle = \frac{4\pi^3 + 18\pi^2 + 54\pi + 81\left(1 - \exp(2\pi/3)\right)}{2\pi^2 \left(2\pi + 3\left(1 - \exp(2\pi/3)\right)\right)}. \tag{8.95}$$

Finally for the ultimate volume crystallinity

$$v_{cs}^\infty = \frac{1}{1 + \langle h_a(u) \rangle} = 0.74 \tag{8.96}$$

is obtained – a value found in the simulation that is in very good agreement with the theoretical Rényi limit according to Eq. (8.92).

**Scattering Data of the Iterated Stochastic Structure.** The computer simulation of the pure stochastic structure evolution process even yields the respective IDF and the scattering data [184]. Here it becomes clear that a standard concept of arranged but distorted structure, the convolution polynomial, is not applicable to

---

[107]If the crystallite requires a space of 1, this interval is $0 < r_3 < 1$.

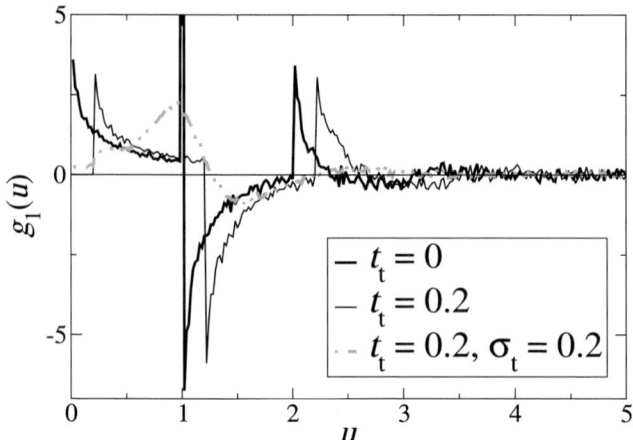

**Figure 8.40.** Computer-simulated IDFs $g_1(u)$ of 1D two-phase structure formed by the iterated stochastic structure formation process. $t_t$ is the thickness of the transition layer at the phase boundary. $\sigma_t$ is the standard deviation of a Gaussian crystallite thickness distribution

stochastic structures. The convolution polynomial will be introduced in Sect. 8.7.3, where we discuss structures that rest on arrangement of particles.

Several computed IDFs of iterated stochastic structures are presented in Fig. 8.40. As long as the crystallite thickness is uniform, the truncated exponentials of the amorphous thickness distributions are clearly identified in the IDF.

If we went back to a diluted structure as discussed in Sect. 8.7.1, one would not observe the exponentials, but only the $\delta$-function of the crystallite size distribution at $u = 1$. The additional terms in $g_1(u)$ that cause the discrete liquid scattering are merely a result of crowding [184, 211]. Let us call these terms packing correlation [184]. Although the packing correlation looks quite strange, it fulfills the zero-sum rule. The packing correlation is of short range. It has vanished for $u > 3 + t_t$. This fact demonstrates that it only comprises correlation among next neighbors. Thus in the IDF or the CDF there is some chance to detect the purely statistical process by visual inspection. In a scattering curve the packing correlation results in a long-period peak looking very similar to that of every other SAXS curve from soft materials. A selection of projected scattering curves is presented in Fig. 8.41. LORENTZ-corrected curves from isotropic materials look very similar. We observe that for nearly uniform thickness of the crystallites the *minima* of the scattering curve are characteristic of the crystallite size. The main maximum (long period) is simply a result of RÉNYI's limit. As already noticed by HOSEMANN [211], a long-period reflection is not conclusive for an arrangement of domains in the material. Long periods are as well found in stochastic structures. Here they simply mark the transition from a diluted to a concentrated system.

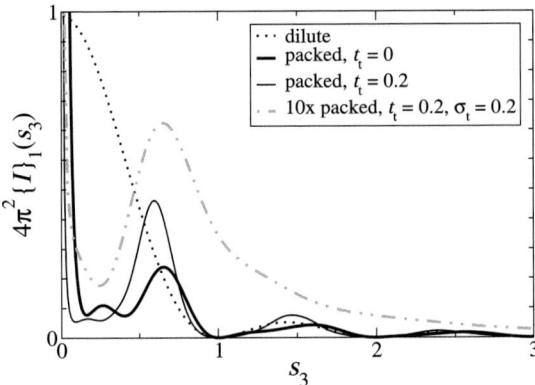

**Figure 8.41.** Liquid scattering projected on the meridian, $\{I\}_1(s_3)$, originating from random placement of lamellae along a shish. The form factor of the dilute system is displayed for reference. Structure parameters as in Fig. 8.40. Uniform crystallite thickness leads to characteristic *minima* in the scattering curves. Note the varying strength of the oscillations

### 8.7.3 Distorted Structure by Infinite 1D Arrangement

**Introduction.** In this section we describe the most simple 1D ordering mechanisms in a distorted structure. There is a common principle in these mechanisms which can be described as follows: imagine you have two bags. One is full of white rods, the other full of black ones. The rod lengths are varying (polydispersity). Rods are randomly taken from the bags and arranged along a line in order to form the structure. There are different "rules of placement" (cf. Fig. 8.42) which are related to different structure evolution mechanisms and which lead to different scattering curves.

**History.** Starting from the 1D point statistics of ZERNIKE and PRINS [116] J. J. HERMANS [128] designs various 1D statistics of black and white rods. He applies these models to the SAXS curves of cellulose. Polydispersity of rod lengths is introduced by distribution functions, $h_i(x)$[108]. HERMANS describes the loss of correlation along the series of rods by a "convolution polynomial". One of HERMANS' *lattice* statistics is named *paracrystal* by HOSEMANN [5, 117]. HOSEMANN shows that the field of distorted structure is concisely treated by the methods of *complex analysis*. A controversial subject is HOSEMANN's extension of 1D statistics to 3D [63, 131, 227, 228].

#### 8.7.3.1 Construction of a 1D Paracrystal

One of two bags contains white rods of differing lengths – the lattice constants described by a length distribution, $h_L(x)$. The average rod length is $\bar{L}$. We randomly

---

[108]The index $i$ discriminates the type ("color") of the rod.

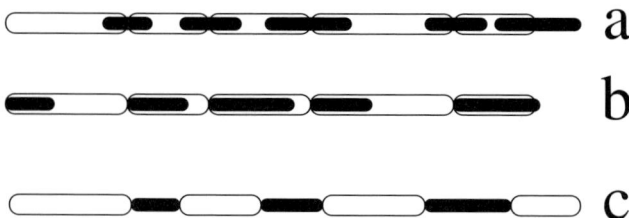

**Figure 8.42.** 1D structural models with inherent loss of long-range order. (a) Paracrystalline lattice after HOSEMANN. The lattice constants (white rods) are *decorated* by *centered* placement of "crystalline" domains (black rods). (b) Lattice model with left-justified decoration. (c) Stacking model with formal equivalence of both phases (no decoration principle)

take white rods from the bag and put them one after the other. This is the sequence of the lattice constants[109] (Fig. 8.42a,b).

**The Convolution Polynomial.** Deliberately selecting the beginning of one of the (white) rods, the *average* distance to the beginning of its neighbor is the long period $\bar{L}$. The *probability* really to find the beginning of the adjacent rod in a distance $x$ is just the distribution of rods in the bag, $h_L(x)$. We consider the *first* of the added rods. Meaning, we have randomly selected a $x'$, for which $h_L(x') \neq 0$ is valid. Then we find the end of the *second* rod with a probability of $h_L(x - x')$ – at the position of the right-shifted (by $x'$) distribution function. Nevertheless, we must not forget to multiply with $h_L(x')$, i.e., with the probability to find the end of the first rod. Finally, the total probability to find the end of the second rod is obtained by integration over all possible choices $x'$ of the first rod

$$h_{2L}(x) = \int h_L(x') \, h_L(x - x') \, dx' \tag{8.97}$$

$$= h_L(x) \star h_L(x). \tag{8.98}$$

This is the definition of a convolution (p. 16, Eq. (2.17)) of the distribution $h_L(x)$ with itself. Repeated induction yields the relation

$$h_{nL} = h_L^{\star n}(x). \tag{8.99}$$

Thus the distance to the end of the $n$-th rod is obtained by $n$-fold convolution of the rod length distribution. A typical series of such lattice constant distributions is demonstrated in Fig. 8.43. Its sum is named *convolution polynomial*.

### 8.7.3.2 Application

**Inconsistency of all Lattice Models.** The decoration principle of all lattice models coupled to polydispersity leads to a fundamental inconsistency of all lattice

---

[109] A decision on the decoration of the lattice has not yet been made. Two variants are depicted in Fig. 8.42a and Fig. 8.42b, respectively. Thus there are different variants of lattice models.

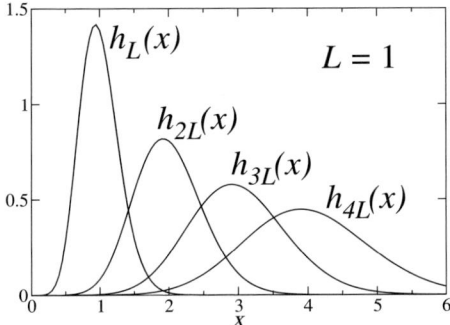

**Figure 8.43.** In all 1D lattice models (including the paracrystal) the higher length distributions of lattice constants are formed by repeated convolution of the fundamental distribution $h_L(x)$

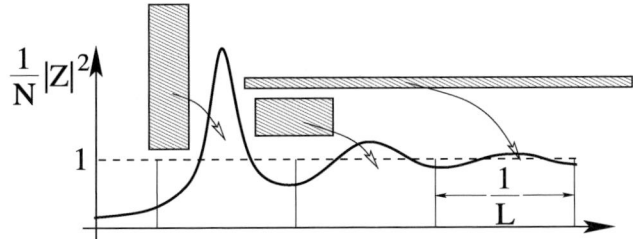

**Figure 8.44.** Effect of "paracrystalline" distortions on a series of reflections in a scattering diagram after compensation of the decay according to POROD's law (lattice factor $(1/N)|Z|^2$). The quadratic increase of integral breadths of the reflections is indicated by boxes of equal area and increasing integral breadth. $L$ is the average long period

models with short-range order (Fig. 8.42a,b). If such models are chosen and the lattice distortion is strong, adjacent decorating particles (i.e., crystalline lamellae) are rather frequently subjected to mutual penetration. These nonsensical artifacts pose a major problem in soft matter science, as here the rod length distributions are rather broad.

The stacking model (Fig. 8.42c) does not carry this inconsistency [128, 229]. It cannot be discriminated from the lattice models if the polydispersity is strong. For small polydispersity even the lattice models make physical sense, because then the mutual penetration is negligible. Computation and fitting of stacking and lattice models are described in Sects. 8.7.3.4 and 8.7.3.3.

### Effect of the 2nd-Order Lattice Distortions on the Scattering Pattern.
As shown by STROBL [230], the integral breadths $B$ in a series of reflections is increasing quadratically if (1) the structure evolution mechanism leads to a convolution polynomial, (2) the polydispersity remains moderate, (3) the "rod-length" distributions can be modeled by Gaussians (cf. Fig. 8.44). For the integral breadth it follows

$$B = \frac{\pi^2 \sigma_L^2}{\bar{L}} s^2, \tag{8.100}$$

with the average long period $\bar{L}$ and the variance $\sigma_L^2$ of the Gaussian distribution of lattice constants.

### 8.7.3.3 The Stacking Model

**Properties and Application.**   The two independent statistical distributions of the two-phase stacking model are the distributions of "amorphous" and "crystalline" thicknesses, $h_1(x)$ and $h_2(x)$. Both distributions are homologous. The stacking model is commutative and consistent. If the structural entity (i.e., the stack as a whole) is found to show medium or even long-ranging order, the lattice model and its variants should be tested, in addition. As a result the structure and its evolution mechanism may more clearly be discriminated.

**Model Construction.**   In the stacking model alternating amorphous and crystalline layers are stacked. Likewise the combined thicknesses in the convolution polynomial are generated by alternating convolution from the independent distributions: $h_3 = h_1 \star h_2$, $h_4 = h_3 \star h_1$, and $h_5 = h_3 \star h_2$. In general it follows

$$h_i = (h_1 \star h_2)^{\star n} \star h_m \tag{8.101}$$

with $n = \text{int}(i/3)$ and $m = \text{mod}(i,3)$. As common in most programming languages, the functions $\text{int}()$ and $\text{mod}()$ designate the integer fraction and the remainder, respectively, of an integer division[110].

If structure visualization by means of the IDF or CDF has shown that $h_1$ and $h_2$ can be modeled by Gaussians, all the combined thickness distributions are Gaussians as well. Each normalized Gaussian is completely described by mean $\bar{d}_i$ and standard deviation $\sigma_i$, and Eq. (8.101) is reduced to a relation

$$\sigma_i^2 = \left( \sigma_1^2 + \sigma_2^2 \right) n + \sigma_m^2 \tag{8.102}$$

for the computation of combined variances. For the mean values, $\bar{d}_i$, of the distributions the obvious law of addition[111] is valid. Thus the model is described completely and can be turned into a computer program. In practice, the correlation among the particles is vanishing rapidly. Thus it is rarely necessary to program an infinite stack. The maximum number of particles required is readily assessed by the number of minima (long-period peaks) that are clearly observable in the IDF or the CDF. Nevertheless, it is important not to violate the zero-sum rule of correlation (p. 158, Fig. 8.30). This means, that a finite model for a structural entity of $m$ correlated particles must contain $3m + 1$ thickness distributions.

---

[110]In the example: Division of $i$ by 3.
[111]For instance: $L = \bar{d}_3 = \bar{d}_1 + \bar{d}_2$, which means: long period = crystalline thickness + amorphous thickness.

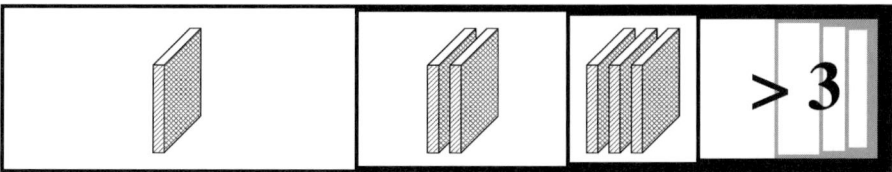

**Figure 8.45.** Sketch of a continued-fraction model for a polyethylene sample during isothermal crystallization and a coupling factor $c = 0.4$. The widths of the nested boxes are proportional to the numbers of crystalline lamellae showing the indicated degree of agglomeration (from left: solos, duos, trios, ...)

**Using the Results of a Fit.** If a fit of the stacking model can be accepted and the range of order is longer than next-neighbor correlation, this finding is indicative for a group of *structure evolution processes*. One may imagine spherulitic growth with a "crystallization" front running through the material while generating particles. Also consistent with the stacking model is the concept of a spinodal decomposition.

The *identification* of the crystalline thickness may be possible from the result of the fit. As a frequent result of a model fit, the relative standard deviations of the fundamental domain thickness distributions, $\sigma_1/\bar{d}_1$ and $\sigma_2/\bar{d}_2$, differ considerably. In many cases the broader distribution can be attributed to the amorphous (or soft) phase. Even higher significance of the assignment can be achieved if the material is studied in time-resolved SAXS experiments during processing (under thermal load, mechanical load). Thus it is not always necessary to resort to secondary methods[112] in order to resolve the ambiguity inherent to Babinet's theorem.

It is not unusual that the fitting of models yields indications for a *complex structure* of the material. Two components of stacks with distinguishable different domain thicknesses may be present and even worse. In general such finding leads to a loss of significance concerning the results of the quantitative analysis. The first-mentioned case can frequently be resolved without loss of significance if it is possible to study highly oriented structures. A second effect that makes structure complex is *polydispersity of the structural entities* as a whole. Because polydispersity is governed by the mathematical principle of the Mellin convolution [2], such structures can be resolved using *modified 1D models* [125, 126].

A third effect that has been encountered [185] is a structure that is made from a *mixture* of different kinds of clusters which are all finite stacks and vary by the number of their members. At the first glance such a material looks as if the *zero-sum rule* were violated – but for each individual cluster it is not. Figure 8.45 shows a sketch of the probability of different clusters in such a structure. In the study [185] the parallel fit using three finite stacks (solos, duos, trios of lamellae) yielded a peculiar coupling of cluster fractions according to the relation

$$n_m = c\,(1-c)^{m-1}, \quad c < 1. \tag{8.103}$$

---

[112]The most simple secondary method is the determination of crystallinity from the density of the material.

which couples the number fractions, $n_m$, of clusters with $m$ members by a single coupling factor $c$. By means of this relation the series of 1D models can, again, be unified in a continued-fraction model. For a value of, e.g., $c = 0.4$ the number fraction of solos is $n_1 = 0.4$. The other 60% of lamellae are agglomerated. From this correlated fraction, again 40% are organized in duos only ($n_2 = 0.24$), the other 60% enjoy a higher degree of agglomeration. Thus for the trios we obtain $n_3 = 0.144$. Continued to infinity, the series of cluster weight parameters is intrinsically computed from a general weight parameter (number of lamellae in the sample) and the coupling parameter $c$.

**Analytical Expressions for Stacks of Infinite Height.**   An analytical expression for the 1D scattering intensity of the ideal stack of infinite height is known since HERMANS [128]

$$I_1(s) = \frac{A_{P_1}}{s^2} \, \Re \left[ \frac{(1 - H_1(s))(1 - H_2(s))}{1 - H_1(s)H_2(s)} \right].$$   (8.104)

$H_1(s) = \mathscr{F}_1(h_1(x))$ and $H_2(s) = \mathscr{F}_1(h_2(x))$ are 1D Fourier transforms of the domain thickness distributions. In general the

$$H_k(s) = a(s)\exp\left(2\pi i \bar{d}_k s\right)$$

are the product of the harmonic function $\exp\left(2\pi i \bar{d}_k s\right)$ carrying the average distance, $\bar{d}_k$, between two phase boundaries, and an attenuation term, $a(s)$, describing the shape of the domain thickness distribution. In particular, if the distributions are able to be approximated by Gaussians we have

$$H_k(s) = \exp\left(-2\pi^2 \sigma_k^2 s^2\right) \exp\left(2\pi i \bar{d}_k s\right)$$   (8.105)

and if the $h_k(x)$ are gamma distributions[113], the solution is according to RULAND [84]

$$H_k(s) = \left(1 - 2\pi i s \frac{\sigma_k^2}{\bar{d}_k}\right)^{\frac{\sigma_k^2}{\bar{d}_k^2}}.$$   (8.106)

**The General Series Expansion for Stacks.**   In practice, 1D scattering intensities can always be modeled by programming the obvious series expansion of such a structural entity: every correlated distance along the stack axis is producing an attenuated oscillation according to $H_k(s)$ that is weighted by the probability of its occurrence under consideration of the zero-sum rule and the related correlation-construction principle (p. 158, Fig. 8.30)

$$I_1(s) = \frac{A_{P_1}}{s^2}\left(1 - \sum_k c_k H_k(s)\right).$$   (8.107)

The constants $c_i$ are balanced to fulfill the zero-sum rule.

---

[113]Gamma distribution: $h(a,v,x) = [a^v/\Gamma(v)]\, x^{v-1}\exp(-ax)$. Most frequently assumed $a = v$. [126]

**Analytical Expressions for Stacks of Finite Height.** By virtue of the just mentioned general series expansion for stacks, even for structural entities built from a finite number of particles analytical solutions can be derived. For a structural entity from $N$ particles of phase 1 the thickness distributions which are the components of the IDF are arranged

$$
\begin{aligned}
N g_1 = {} & N h_1 \\
& + (N-1)\left(h_2 - 2h_1 \star h_2 + h_1 \star h_2 \star h_1\right) \\
& + (N-2)\left(h_2 - 2h_1 \star h_2 + h_1 \star h_2 \star h_1\right) \star h_1 \star h_2 \\
& \vdots
\end{aligned}
\tag{8.108}
$$

in zero-sum groups and the scattering intensity is computed by Fourier transformation. In the presented equation the arguments of the functions have been left out for clarity. The complete deduction has been published by RULAND [84]. Its solution is

$$
I_1(s) = \frac{A_{P_1}}{s^2} \, \Re \left[ \frac{(1-H_1)(1-H_2)}{1-H_1 H_2} + \frac{H_2 (1-H_1)^2 \left(1-(H_1 H_2)^N\right)}{N (1-H_1 H_2)^2} \right].
\tag{8.109}
$$

If the structural entities contain varying numbers of particles (solos, duos, trios, ...), RULAND [84] deduces

$$
I_1(s) = \frac{A_{P_1}}{s^2} \left\{ \Re \left[ \frac{(1-H_1)(1-H_2)}{1-H_1 H_2} + \frac{H_2 (1-H_1)^2 \left(1-\left\langle (H_1 H_2)^N \right\rangle\right)}{\langle N \rangle (1-H_1 H_2)^2} \right] - J_0 \right\}
\tag{8.110}
$$

with $\langle N \rangle$ denoting the number-average number of particles per structural entity. Here RULAND annotates that it makes a difference, whether the structural entities are surrounded by "amorphous matrix material" ($J_0 = 0$), or whether they abut upon each other. In the latter case, the form factor of the average structural entity

$$
J_0(s) = \left[ \operatorname{Im} H_1 \left(\frac{s}{2}\right) \right]^2 \frac{1 - \langle \cos(2\pi \bar{L} s) \rangle}{\pi^2 \bar{L}^2 s^2 \langle N \rangle}
$$

must be subtracted[114] [84]. In the equation $\bar{L} = \bar{d}_1 + \bar{d}_2$ is the average long period of the stack. BURGER [231] makes aware of the fact that in most of the practical cases $J_0$ may be dropped, because it is, in general, only significant at small $s$ where a divergence $s^{-2}$ in the 3D scattering pattern is dominating the shape of the scattering curve.

---

[114]At a first glance this subtraction appears to be a violation of the zero-sum rule. However, here an exception has to be made, because particles merge upon direct contact of adjacent structural entities, and thus the number of particles is reduced – just by the amount deduced by RULAND.

**Practical Value.** The presented analytical expressions are very useful, predominantly for the analysis of the scattering from *weakly distorted nanostructures*. Because of their detailed SAXS curves, direct fits to the measured data return highly significant results (cf. Sect. 8.8.3). Nevertheless, some important corrections have to be applied [84]. They comprise deviations from the ideal multiphase structure as well as thorough consideration of the setup geometry and machine background correction (cf. Sect. 8.8).

### 8.7.3.4 The Lattice Model

**Properties.** There is only one stacking model, but several variants of lattice models. The two statistically independent thickness distributions of a short-range lattice are the distribution of lattice constants, $h_{\tilde{L}}(x)$, and the thickness distribution of the decorating ("crystalline") domains, $h_1(x)$. In general, even the decorating particles are not identical but subjected to polydispersity. In analogy to crystallography this variation is addressed as substitutional disorder. Lattice models are not commutative: upon exchange of $h_1(x)$ for $h_2(x)$ the shape of the IDF and the scattering curve is changed. The computational effort of testing a lattice model is thus twice the effort of testing the stacking model: after a regression has converged, amorphous and crystalline thickness must be exchanged and another series of regression runs must be performed starting from the exchanged parameter set. The best fit indicates which thickness is probably the decorating phase. As already mentioned, lattice models are only consistent for weakly distorted structure (cf. Sect. 8.7.3.2).

**Model Construction and Structure Evolution Processes.** The principle of how to construct a distorted lattice has already been introduced in Sect. 8.7.3.1. Here it is refined and related to structure evolution. In the ultimate topology the lattice constants of a paracrystal (white rods in Fig. 8.43a) can not directly be found. For this purpose we call them *pseudo long-periods*, $\tilde{L}$, and mark them by a tilde. Thus the fundamental convolution polynomial is generated from multiple convolutions of $h_{\tilde{L}}$ (cf. Fig. 8.43). A notion behind the lattice model is structure evolution according to a nucleation-and-growth mechanism. Nuclei are first generated along a line. This primary structure is called a row structure. In a second step the nuclei grow to form the domains (lamellae). Only then it must be decided how to decorate the lattice. Frequently it is assumed that the decorating particle is *centered* on the lattice point, but also *oriented particle growth* "always in the same direction" (Fig. 8.43b) or "randomly to the right or left" are possible. Each different evolution mechanism leads to different structural entities, which can be discriminated if the distortion of the lattice is weak.

Let us demonstrate the centrally-decorated lattice model[115]. The corresponding geometrical consideration for the deduction of the higher thickness distributions is somewhat involved. Figure 8.46 presents a geometrical construction scheme that helps to link both the observable distribution of long periods, $h_L(x)$, and the observ-

---

[115]The other mentioned variants are easier to handle.

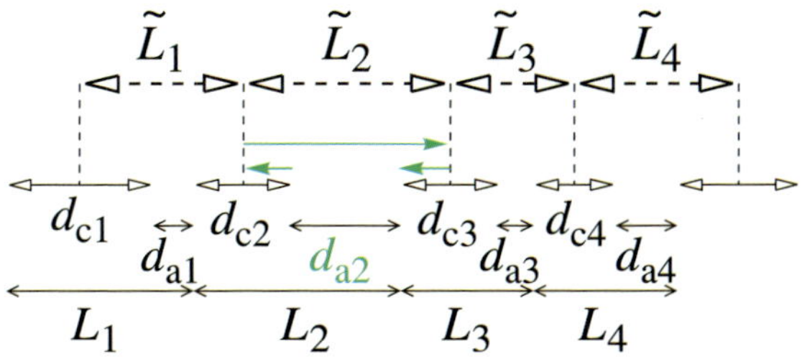

**Figure 8.46.** Construction of observable distances $(d_{ai}, L_i)$ in a paracrystal from its fundamental distances, $\tilde{L}_i$ (not observable pseudo long-periods), and $d_{ci}$ (observable crystallite thicknesses). Advancing to the bottom the relations to the observable thicknesses of amorphous layers, $d_{ai}$, and the long periods, $L_i$, are sketched

able distribution of amorphous thicknesses, $h_a(x)$, to the generating distributions[116] of the lattice model, $h_c(x)$ and $h_{\tilde{L}}(x)$. For the example of $d_{a2}$ the sketch demonstrates (by thin horizontal arrows with filled heads), how the amorphous thickness is reduced to the pseudo long-periods and *halves* of the crystallite thickness distributions. If the distributions can be approximated by Gaussians we thus obtain for the variances of the dependent thickness distributions

$$\sigma_i^2 = n\,\sigma_{\tilde{L}}^2 + \frac{\sigma_1^2}{2} \tag{8.111}$$

with $n = \mathrm{int}\,((i+1)/3)$ and $m = \mathrm{mod}\,((i+1),3) - 1$. Equation (8.111), Fig. 8.46 and the zero-sum rule is sufficient information to write a computer program that models the paracrystalline lattice model for a structural entity built from a finite number of "crystallites".

**Analytical Expressions for Lattice Models.**  Concerning the aforementioned paracrystalline lattice, an analytical equation has first been deduced by HERMANS [128]. His equation is valid for infinite extension. RULAND [84] has generalized the result for several cases of finite structural entities. He shows that a master equation

$$I_1(s) = \frac{A_{P_1}}{s^2}\left(1 - \Re H_1(s) + 2\left(\Im H_1\left(\frac{s}{2}\right)\right)^2\left(\frac{1}{N}|Z(s)|^2 - 1\right)\right) \tag{8.112}$$

describes both infinite and finite paracrystalline lattices for different cases. The variants are discriminated by different lattice factors, $|Z(s)|^2/N$. With the lattice factor

$$\frac{1}{N}|Z(s)|^2 = \Re\left(\frac{1 + H_{\tilde{L}}(s)}{1 - H_{\tilde{L}}(s)}\right), \tag{8.113}$$

---

[116]In the sketch the fundamental distances carry arrow heads with open triangles. The arrows of observable distances are solid-line double-head arrows.

HERMANS' equation for the *infinitely* extended lattice is obtained. For a material built from *finite* structural entities containing an average of $\langle N \rangle$ particles RULAND obtains

$$\frac{1}{N} |Z(s)|^2 = \text{Re} \left( \frac{1 + H_{\bar{L}}(s)}{1 - H_{\bar{L}}(s)} - \frac{2 H_{\bar{L}}(s) \left( 1 - \langle H_{\bar{L}}^N(s) \rangle \right)}{\langle N \rangle \left( 1 - H_{\bar{L}}(s) \right)^2} \right) - J_0(s). \quad (8.114)$$

In analogy to the treatment of the stacking model $J_0(s) = 0$ is valid, if the structural entities are embedded in matrix material. Compact material, again, may require a correction because of the merging of particles from abutting structural entities

$$J_0 = \frac{1 - \left\langle \cos \left( 2\pi \bar{\bar{L}} N s \right) \right\rangle}{2\pi^2 s^2 \bar{\bar{L}}^2 \langle N \rangle}. \quad (8.115)$$

Independently, BURGER [231] develops analytical equations for lattice models *without* substitutional disorder. His results are special cases of the models presented by RULAND.

As has already been mentioned in the discussion of the stacking model, such equations are particularly useful for the analysis of nanostructured material with weak disorder in order both to assess the perfection of the material and to discriminate among lattice and stacking models (cf. Sect. 8.8.3).

### 8.7.3.5 Model Fitting: Choice of Starting Values for the Model Parameters

Stacking model and lattice model have the same number of model parameters. A global weight parameter $W = A_{P_1}/2$ adjusts the integral[117] of the IDF. Because $A_{P_1} = -G_1(0)$ has already been determined in the pre-evaluation, a suitable starting value for the regression program is at hand. The next two parameters are the average thicknesses of the phases (for the stacking model) or the long period and the thickness of the decorating phase (for the lattice model). The range of suitable starting values is obtained from the long period as determined from the minimum of the IDF. In the model function all the standard deviations are expressed by relative standard deviations, e.g., $\sigma_1/\bar{d}_1$ and $\sigma_2/\bar{d}_2$ for the sake of steady convergence. Choose a starting value of 0.3 for all the relative standard deviations. Narrow values are not easily broadened by the regression program. Analogously, the program will not be able to escape from distributions that are too broad and flat. For more information concerning model fitting by regression consult Chap. 11.

---

[117]In practical application only the positive branch of the IDF will be fitted. This explains the halving of the weight.

## 8.8 Nanostructures – Soft Materials with Long Range Order

**Perfection of Structure in Nanostructured Materials.**   An aim of modern nanotechnology is the fabrication of materials with highly perfect structure on the nanometer scale. The distortion of such nanostructured materials can be studied by SAXS methods. Frequently the material is supplied as a very thin film with predominantly uniaxial correlation among the nanodomains. Under these constraints the nanodomains are frequently arranged in such a way that the normal to the film is a symmetry axis: rotation of the film on the sample table does not change the scattering (fiber symmetry).

In this case a suitable *setup geometry* for the beamline is SAXS in symmetrical-reflection geometry (SRSAXS). Sometimes an investigation with the beam imping-ing under fixed grazing incidence may as well be suitable. Moreover, it is wise to assess the orientation distribution $g(\phi)$ of the structural entities in a similar manner as in a classical texture goniometer: SAXS is measured as a function of sample tilt at a fixed scattering angle $2\theta$.

SRSAXS is utilized, because the scattering curve measured in symmetrical re-flection is readily transformed into a 1D scattering curve, $I_1(s_3)$. The importance of such curves for the analysis of structure has been demonstrated in the preceding sections of this chapter.

### 8.8.1 Required Corrections of the Scattering Intensity

Special care has to be taken concerning data pre-evaluation if the scattering of highly oriented nanostructures are investigated in symmetrical reflection or at grazing inci-dence. Absorption correction is delicate (cf. Sect. 7.6.3). Even a refraction correction (Sect. 7.6.5) may be necessary[118].

Moreover, if the experiment is set up in such a way that either the angle of in-cidence, $\alpha_i$, or the angle of exit, $\alpha_e$, are close to the critical angle, $\theta_c$, the classical scattering theory ("*Born approximation*") is no longer valid as far as it states that the incident wave field is the effective field at each point of the scatterer. In this case the viable theoretical approach is to apply a first-order perturbation of the incident wave, which is induced by the structure of the scatterer itself. This method is called the *distorted-wave Born approximation* (DWBA) [232–235]. A review of this approach, the presentation of a computer program for data analysis, and demonstrations of the effect of wave distortion has been published by LAZZARI [236]. In the field of soft matter the DWBA-approach has recently been utilized by the Korean group of REE [237–239]. Discussion of the corresponding data treatment is beyond the scope of this book.

RULAND and SMARSLY [9, 84, 240] can analyze their recorded data in the clas-sical Born approximation, but have to correct for the special geometry of the grazing incidence experiment. They propose not to carry out the necessary corrections in a

---

[118]A refraction correction should be taken into account, if the experiments are carried out with a USAXS setup. Then the lattice constants of the investigated nanostructures are typically above 100 nm.

separate pre-evaluation, but to include correction terms in the analytical model functions. They directly fit the measured scattering curve by a *compound model* [84] like

$$I_{mod}(s) = w A_{sr}(s) \left[ I_{sr}(s) H_z^2(s) + I_{Fl} \right]. \tag{8.116}$$

Here $w$ is a weighting factor. $A_{sr}(s)$ is the absorption factor (in this case for symmetrical absorption), $H_z(s)$ and $I_{Fl}$ consider the non-ideal character of the two-phase topology (cf. p. 124, Fig. 8.10) by consideration of a smooth phase transition zone and density fluctuations inside the phases.

The *analytical structural model* for the topology of the nanostructure is defined in $I_{sr}(s)$. For many imaginable topologies such models can be derived by application of scattering theory. Several publications consider layer topologies [9, 84, 231] and structural entities built from cylindrical particles [240, 241]. In the following sections let us demonstrate the principle procedure by means of a typical study [84].

### 8.8.2 $I_1(s)$ from a Nanostructured Layer System

Let us consider a nanostructured thin film built from lamellar particles [84]. If the principal axis of layer stacks is oriented normal to the film surface, the scattered intensity measured in symmetrical-reflection geometry (SRSAXS) is

$$I_{sr}(s) = I_{ori}(s, \phi = 0) \propto \frac{g(\phi = 0)}{s^2} I_1(s) \tag{8.117}$$

proportional to the isotropic[119] intensity. This equation is only valid for ideal point-focus. In practice, the primary beam has a finite extension. Important for the smearing of the observed curve is the extension of the primary beam in the direction perpendicular to the plane of incidence. The respective profile, $W(y)$, should be known. In this case the smearing is described by the equation[120]

$$I_{sr}(s) \propto \int_0^\infty W(y) \frac{g[\arctan(y/s)]}{s^2 + y^2} I_1 \left( \sqrt{s^2 + y^2} \right) dy.$$

In the standard setup $W(y)$ is the profile of the primary beam in horizontal direction. In order to solve the smearing integral, the orientation distribution of the layer normals, $g(\phi)$, is approximated by a Poisson kernel[121] and $W(y)$ is approximated by a shape function with the integral breadth $2 y_{max}$ of the primary beam perpendicular to the plane of incidence. In the simplified result

$$I_{sr}(s) \propto \int_0^{y_{max}} \frac{1 - q^2}{s^2 (1-q)^2 + y^2 (1+q)^2} I_1 \left( \sqrt{s^2 + y^2} \right) dy, \tag{8.118}$$

the Poisson parameter $q$ describes the possible orientation of the material from the isotropic orientation of layer stacks ($q = 0$) to perfect uniaxial orientation ($q = 1$). If the orientation is high, the approximation

---

[119]This is in 3D the solid-angle averaged intensity of the stack – as if an isotropic material (made from stacks of layers) had been studied in a conventional setup.

[120]The same equation is considered for the smearing of the classical Kratky camera.

[121]The Poisson kernel is the Lorentzian on the orientation sphere (cf. Sect. 9.8).

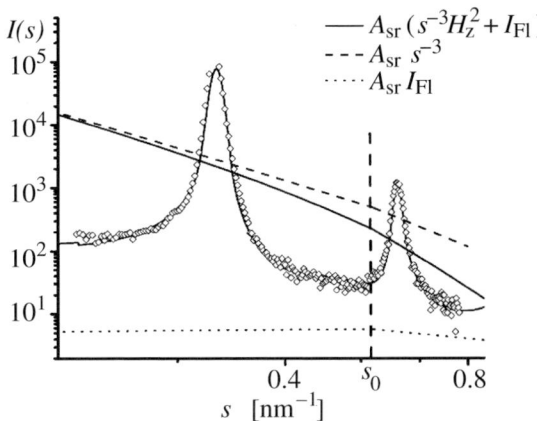

**Figure 8.47.** SRSAXS raw data (*open symbols*) and model fit (*solid line*) for a nanostructured material using a finite lattice model. The model components are demonstrated: absorption factor $A_{sr}$, density fluctuation background $I_{Fl}$, smooth phase transition $H_z$. The solid monotonous line demonstrates the shape of the Porod law in the raw data. At $s_0$ the absorption is switching from fully illuminated sample to partial illumination of the sample

$$I_{sr}(s) \propto \frac{1}{s} \arctan\left(\frac{y_{max}}{s} \frac{1+q}{1-q}\right) I_1(s) \tag{8.119}$$

is valid. For very high orientation ($0.95 < q < 1$) and small $s$–values even

$$I_{sr}(s) \propto \frac{1}{s} I_1(s) \tag{8.120}$$

is valid – similar to respective results for measurement with a Kratky camera.

By trial-and-error it is possible to find out, which of the successive approximations is valid: $y_{max}$ can be measured or assessed from the beamline geometry. Together with $q$ it can be varied within reasonable intervals, in order to fit analytical models for $I_1(s)$ (e.g., after Eq. (8.110) or Eq. (8.112)) to measured data.

### 8.8.3 Typical Results

RULAND and SMARSLY [84] study silica/organic nanocomposite films and elucidate their lamellar nanostructure. Figure 8.47 demonstrates the model fit and the components of the model. The parameters $I_{Fl}$ and $\sigma_z$ (inside $H_z^2$) account for deviations from the ideal two-phase system. $A_{sr}$ is the absorption factor for the experiment carried out in SRSAXS geometry. In the raw data an upturn at $s_0$ is clearly visible. This is no structural feature. Instead, the absorption factor is changing from full to partial illumination of the sample. For materials with much stronger lattice distortions one would mainly observe the Porod law, instead – and observe a sharp bend – which are no structural feature, either.

Figure 8.48 visualizes the *selectivity* of this RULAND-SMARSLY method for

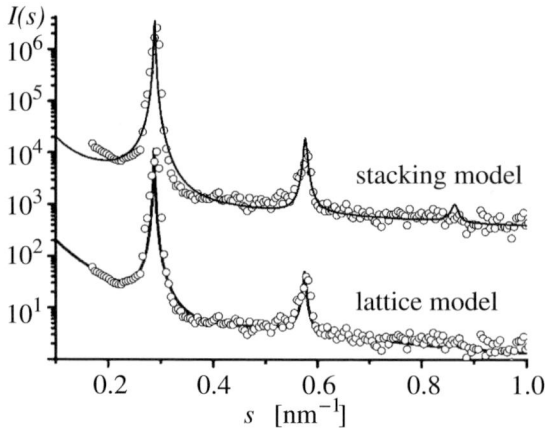

**Figure 8.48.** Best fits of stacking model and lattice model to the data from Fig. 8.48. The lattice model fits much better. Data sets are shifted for clarity

weakly distorted nanostructures and differing structure evolution mechanisms. The best fit of the lattice model to the scattering data is significantly better than the best fit of the stacking model.

**Disorder of Nanocomposites and Common Polymers.** If one compares the distortion parameters of particular nanocomposites with those of common polymer materials, the relative standard deviations are generally smaller by 1 order of magnitude. More than 30 layers are correlated to each other, whereas the correlation in commercial polymer materials is generally ranging shorter than 4 layers.

## 8.9 Anomalous X-Ray Scattering

**Application.** Anomalous X-ray diffraction (AXRD), anomalous wide-angle X-ray scattering (AWAXS), and anomalous small-angle X-ray scattering (ASAXS) are scattering methods which are selective to chemical elements. The contrast of the selected element with respect to the other atoms in the material is enhanced. The phase problem of normal X-ray scattering can be resolved, and electron density maps can be computed.

Consider a polystyrene-(b)-polybutadiene star block copolymer with four arms coupled by a central Si-atom. Or consider a metal catalyst (e.g., Au) supported in activated carbon. Then the scattering of only the selected element (Si, Au, respectively) can be extracted [242]. Even the distribution of the elements in the material can be mapped based on ASAXS data. A concise review of the ASAXS method in combination with AXRD and AWAXS has been published by GOERIGK et al. [243].

**Technical Requirements.**   AXS requires an X-ray source with easily tunable, monochromatic photon wavelength. This means that a respective device can only be operated at a synchrotron. In general a 2D detector is used.

**Procedure.**   After the element of interest has been chosen, an X-ray absorption edge of the element that is inside the tuneable range of the synchrotron is selected[122]. In general this is a K-edge or an L-edge. For the conversion between energy, $E$ in keV, and the wavelength of radiation, $\lambda$ in nm, the relation

$$\frac{E}{\text{keV}} = \frac{1.2398\,\text{nm}}{\lambda} \tag{8.121}$$

holds.

Then two or three scattering or diffraction measurements at different wavelengths to the left and to the right of the edge are performed and evaluated. The most simple way is a subtraction of two scattering patterns. After this operation the signal of all normally scattering material is extinguished. *Correlation functions* of only the anomalous scatterers with each other, of only the normal scatterers with each other, and the cross-correlations of anomalous scatterers with normal ones can be computed and result in a detailed view on the size distributions of the anomalous scatterers and their arrangement in the material.

**For Comparison: Notions of Normal Scattering.**   As the electron density is assumed to be a real quantity, it directly follows the central symmetry of scattering patterns known by the name Friedel's law [244]. Friedel pairs are Bragg reflections $hkl$ and $\overline{hkl}$ that are related by central symmetry. Concerning their scattering amplitudes, Friedel pairs have equal amplitude $|A_{hkl}| = |A_{\overline{hkl}}|$ and opposite phase $\phi_{hkl} = -\phi_{\overline{hkl}}$. Consequently, in the scattering intensity the phase information on the structure factor is lost.

**Fundamental Notions of Anomalous Scattering.**   As the wavelength of the X-rays is approaching an absorption edge, the atom can absorb the energy of the photon by lifting an electron to a higher shell. As absorption is, in general[123], changing a real quantity to a complex number, the electron density is no longer a real number in anomalous scattering. Consequently, Friedel's law is broken, and scattering patterns need not be symmetrical any longer (Fig. 8.49). Both the amplitude and the phase of the members of Friedel pairs may be changed. Changing the amplitude can be used for contrast enhancement. Changing the phase can be used for restoration of phase information.

Theoretical considerations of anomalous scattering start from the general relation

$$f(E) = f_0 + f'(E) + i f''(E) \tag{8.122}$$

---

[122]For instance, the K-absorption of Au is at $\lambda_{\text{Au K}}= 0.01536$ nm, the K-absorption of Si is at $\lambda_{\text{Si K}}= 0.67423$ nm.

[123]Other relevant fields in physics are: dielectric loss, mechanical loss modulus

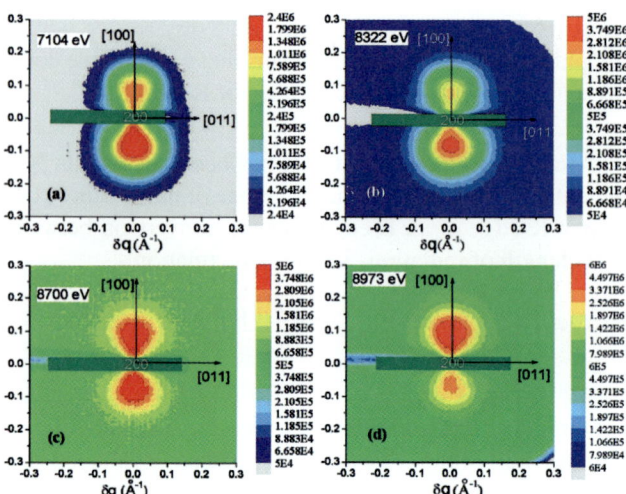

**Figure 8.49.** 2D ASAXS spectra (logarithmic scale) at different photon energies, appearing around a 200 Bragg peak of a Cu-Ni-Fe single crystal aged for 3 h at 773 K. The Bragg peak itself is masked by a beamstop to prevent damage to the detector (a) near the Fe absorption edge, (b) near the Ni absorption edge. (c), (d) near the Cu edge, where both Fe and Ni fluorescence contribute to increase the background. Source: LYON *et al.* [245]

for the atomic scattering factor. For normal scattering it is sufficient to consider the non-resonant term, $f_0 = Z$, which is equal to the number of electrons, $Z$, of the atom[124]. The curve of the imaginary part of anomalous dispersion, $f''(E)$, must be measured by X-ray absorption (EXAFS: Extended X-ray Absorption Fine Structure) in order to determine shape and exact position of the absorption edge in the studied sample. The real part of anomalous dispersion, $f'(E)$, is computed from $f''(E)$ by numerical integration using a Kramers-Kronig relation

$$f'(\omega) = \frac{2}{\pi} \int_0^\infty \frac{\omega' f''(\omega') \, d\omega'}{\omega^2 - \omega'^2}.$$  (8.123)

Since $f'(E)$ and $f''(E)$ are sharply varying at energies within 10 eV of the absorption edge, the monochromatization of the probing X-rays must be very high. Concerning the beam exiting the monochromator, its height and angle must not vary as the wavelength is tuned.

Reviews of scattering methods in materials science that contain examples of the application of anomalous scattering have been published by BALLAUFF [246] and FRATZL [247].

---

[124]This definition leads to an electron density measured in "electrons per nm$^3$" (cf. Sect. 7.10.1). If we are aiming at a treatment in terms of scattering cross-section we define $f_0 = Z r_e$, instead.

# 9 High but Imperfect Orientation

**Introduction.** The following two chapters are devoted to the evaluation of the orientation of structural entities in the studied material, not to the analysis of the inner structure (topology) of these entities. First discussions of the problem of orientation smearing go back to KRATKY [248, 249]. Unfortunately, the corresponding mathematical concepts are quite involved, and a traceable presentation would require mathematical reasoning that is beyond the scope of this textbook. Thus only ideas, results and references are presented.

Whenever we are considering orientation, we are dealing with anisotropic scattering data. Orientation is most frequently analyzed in 2D scattering data with fiber symmetry or in pole-figure data recorded by means of a texture goniometer.

The aim of orientation analysis is not only the quantitative description of orientation, but also the separation of orientational effects from topological ones – ultimately meaning the desmearing of imperfect orientation in order to reconstruct the scattering pattern of the perfectly oriented structural entities.

In the present chapter we assume that rigid and well-defined structural entities are present but not perfectly oriented – similar to a bundle of jackstraws (Fig. 9.1). Moreover, we assume that there is no coupling between the orientation of a struc-

**Figure 9.1.** Imperfect orientation in a bundle of structural entities (jackstraws)

tural entity and its structure. Two general topics are addressed

1. The general relation (in 3D) between the orientation distribution and the scattering of the perfectly oriented structural entity (Sect. 9.2).

2. The use of found Master orientation distributions for the purpose of iterative orientation desmearing in the general case (Sect. 9.3).

Moreover, some issues concerning scattering patterns with fiber symmetry are discussed

1. The use of unimodal meridional reflections for determination of the orientation distribution and desmearing (Sect. 9.6).

2. The influence of finite size and imperfect orientation of the entities on the shape of the reflections. Separation of unimodal orientation distributions by means of RULAND's streak method, and assessment of the analytical shape of the orientation distribution (Sect. 9.7).

3. Detection limits for splitting of orientation distributions (Sect. 9.7).

4. Modeling of orientation distributions by analytical functions and shape change of the latter upon use, i.e., by mapping them on the orientation sphere (Sect. 9.8).

## 9.1 Basic Definitions Concerning Orientation

**Problem.** Let a polymer fiber contain rod-shaped structural entities in an amorphous matrix with some preferential orientation. Let us assume that the rods are crystalline. Our interest is to study the crystalline structure of the rods. Instead of sharp *hkl* reflections we observe that each reflection is smeared over a spherical cap in solid angle. Thus the observed intensity is suitably expressed in polar coordinates

$$I_{hkl}(\mathbf{s}) = I_{hkl}(s, \varphi, \psi). \tag{9.1}$$

The goal is, both to gather information on the orientation distribution $g(\varphi, \psi)$ of the rods, and to reconstruct the scattering intensity $I_{opt}(\mathbf{s})$ of the perfectly oriented rod.

### 9.1.1 Pole Figures and Their Expansion

**Definition: Pole Figure.** A *pole figure*, $g_{hkl}(\varphi, \psi)$ of the reflection $I_{hkl}(\mathbf{s})$, is defined as the *projection* (cf. Fig. 9.2) of the scattering intensity of the reflection

$$g_{hkl}(\varphi, \psi) = \int_{s_{min}}^{s_{max}} I_{hkl}(s, \varphi, \psi) \, ds \tag{9.2}$$

on the *orientation sphere*[1]. With respect to the structure of the crystallites, $g_{hkl}(\varphi, \psi)$ is the *orientation distribution* of the *hkl* netplane normal directions. The range of integration may be reasonably chosen, sometimes even decreased to a thin spherical shell. Properly *normalized*, $g_{N,hkl}(\varphi, \psi)$ describes the probability to find a netplane normal oriented in the direction $(\varphi, \psi)$. This definition[2] of orientation distributions is a suitable starting point for theoretical considerations. In practice, one will often simply consider the orientation distribution of the structural entities, $g(\psi, \varphi)$.

---

[1] The orientation sphere is defined as a sphere of *radius* 1.

[2] Conceded – Crystallography carries out a further projection from the surface of the sphere to a circular disk before it calls the function a pole figure. We omit this distorting stereographic projection. In this case, in fact, a pole figure and an orientation distribution are very similar. If we say "pole figure", we bear in mind a well-separated *reflection* and its image on the orientation sphere. If we speak of an "orientation distribution", we imagine the *structural entities*, instead.

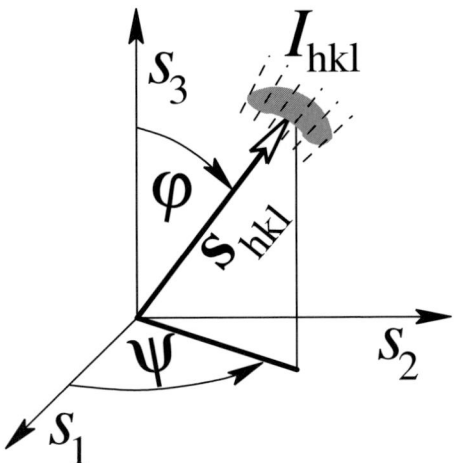

**Figure 9.2.** Reflection $I_{hkl}$ (**s**) smeared by misorientation of the ensemble of crystallites (structural entities). The center of gravity is found at $\mathbf{s}_{hkl}$. Definitions of polar coordinates $\varphi$, $\psi$ are sketched. Dashed lines indicate the *radial* direction of the integration that leads to a projection onto the *orientation sphere* resulting in the pole figure $g_{hkl}(\varphi, \psi)$ of the reflection intensity

**Measurement of Pole Figures.** Pole figures are directly measured in a *texture goniometer*. The geometry of such an instrument is sketched in Fig. 9.3. Before the pole figure measurement starts, the scattering angle, $2\theta_{hkl}$, related to the position of the reflection in reciprocal space, $|\mathbf{s}_{hkl}|$, (cf. Fig. 9.2) is fixed. In the pole-figure scan the sample is – with respect to the dashed vertical line – tilted ($\varphi$) and rotated ($\psi$), respectively.

**Pole Figure and Multipole Expansion.** Pole figures, $g(\varphi, \psi)$, and other functions that are suitably expressed in spherical polar coordinates are favorably expanded

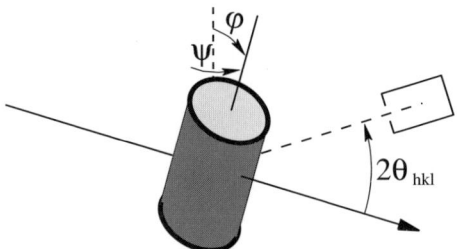

**Figure 9.3.** Scattering geometry for the measurement of pole figures in a texture goniometer. The scattering angle $2\theta_{hkl}$ is fixed. $\varphi$ and $\psi$ are scanned

$$g(\varphi, \psi) = \sum_{\ell=0}^{\infty} \sum_{m=-\ell}^{\ell} a_{\ell}^{m} Y_{\ell}^{m}(\varphi, \psi) \tag{9.3}$$

in spherical harmonics. An expansion in spherical harmonics is called a multipole expansion. The $a_{\ell}^{m}$ are called expansion coefficients. With increasing modulus of their indices, the spherical harmonics, $Y_{\ell}^{m}(\varphi, \psi)$, describe preferential orientation more and more accurately, starting from the isotropic state. The functions themselves

$$Y_{\ell}^{m}(\varphi, \psi) = \begin{cases} P_{\ell}^{m}(\cos\varphi) \cos m\psi & 0 \le m \le \ell \\ P_{\ell}^{|m|}(\cos\varphi) \sin|m|\psi & -\ell \le m < 0 \end{cases} \tag{9.4}$$

are defined by Legendre functions [4], $P_{\ell}^{m}(x)$. The Legendre functions with $m \ne 0$ are called associated Legendre functions. For $m = 0$ the Legendre polynomials

$$P_{\ell}^{0}(x) = P_{\ell}(x) \text{ with}$$
$$\begin{aligned} P_0(x) &= 1, \\ P_1(x) &= x, \\ P_2(x) &= \frac{1}{2}\left(3x^2 - 1\right), \end{aligned} \tag{9.5}$$

$$\vdots$$

are obtained. They are most important for the discussion of orientation.

The expansion coefficients of the spherical harmonics for a deliberate distribution $g(\varphi, \psi)$ are easily computed from

$$a_{\ell}^{m} = \int_{\omega} g(\varphi, \psi) Y_{\ell}^{m}(\varphi, \psi) \, d\omega, \tag{9.6}$$

with the solid-angle element defined by $d\omega = \sin\phi \, d\varphi d\psi$. The reason for this simple computational scheme is the definition of the spherical harmonics – as an orthonormal system of functions.

### 9.1.2 The Uniaxial Orientation Parameter $f_{or}$

The uniaxial orientation parameter is the most simple way to characterize preferred orientation. It is simple, because it is only a number – in fact, $f_{or} = a_2^0$ is the first non-trivial expansion coefficient in a multipole expansion of the normalized

$$\int_{\omega} g_N(\varphi, \psi) \, d\omega = 1 \tag{9.7}$$

pole figure, $g_N(\varphi, \psi)$. $f_{or}$ is also called *Hermans' orientation function* [250].

**Interpretation of $f_{or}$.** For materials that exhibit fiber symmetry (i.e., $g(\varphi, \psi) = g(\varphi)$) induced by structural entities with fiber symmetry

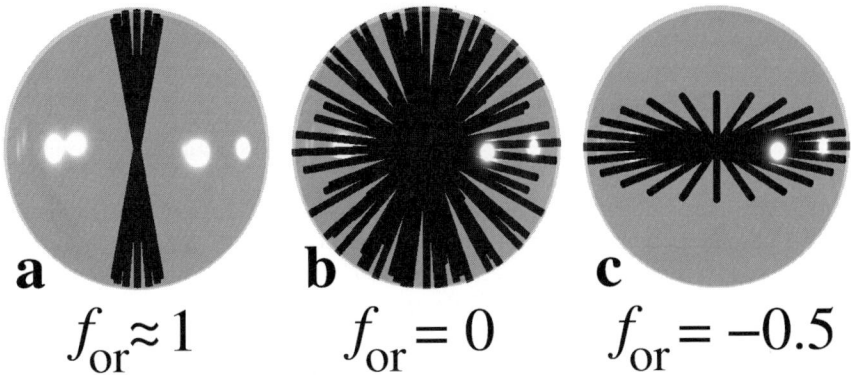

**Figure 9.4.** The orientation of structural entities (rods) in space with respect to the (vertical) principal axis and the values of $f_{or}$, the uniaxial orientation parameter (Hermans' orientation function) for (a) fiber orientation, (b) isotropy, (c) film orientation

themselves[3] (F2-materials), the meaning of $f_{or} \in [-0.5, 1]$ is easily understood (Fig. 9.4). If $\varphi$ is the angle between the principal axis of the material, and $g(\varphi)$ is the orientation distribution of the axes of the structural entities, then $f_{or} = 1$ means that all the entities are perfectly oriented parallel to the fiber direction of the material. This is a typical *fiber orientation*. In the isotropic case the structural entities are oriented at random, and a determination of the orientation parameter yields $f_{or} = 0$. If the principal axes of the structural entities are all oriented in a plane perpendicular to the principal axis of the material (*film orientation*), the orientation parameter returns $f_{or} = -0.5$.

As are the other multipole-expansion coefficients, the uniaxial orientation parameter is computed from Eq. (9.6). For materials with fiber symmetry the relation simplifies[4] and

$$a_\ell = \frac{2\ell + 1}{2} \int_0^\pi g(\varphi) \, P_\ell(\cos\varphi) \sin\varphi \, d\varphi \qquad (9.8)$$

is obtained. Trivial is the coefficient $a_0$. It quantifies the isotropic fraction[5] of structural entities in the material. The coefficient $a_1 = 0$ is vanishing, because all orientation functions are even ($g(s) = g(-s)$). Thus the first non-trivial coefficient in the multipole expansion of the normalized fiber-symmetric pole figure is the uniaxial orientation parameter,

---

[3]Imagine the preferred orientation of a polymer fiber induced by preferred orientation of the polymer chains.

[4]In the literature slightly different definitions of the multipole expansion are found, depending on how the pre-factor $(2\ell + 1)/2$ is distributed between expansion equation and the definition of the coefficients. Cf. (Ward [251], eq. 5.2)

[5]That is, "jackstraws" oriented at random orientation

$$f_{or}\left(g\left(\varphi\right)\right) = a_2\left(g_N\left(\varphi\right)\right)$$

$$= \frac{5}{2}\int_0^{\pi} g_N\left(\varphi\right) P_2\left(\cos\varphi\right)\sin\varphi\,d\varphi$$

$$= \frac{5}{4}\int_0^{\pi} g_N\left(\varphi\right)\left(3\cos^2\varphi - 1\right)\sin\varphi\,d\varphi. \qquad (9.9)$$

In practice, either a pole figure has been measured in a texture-goniometer setup, or a 2D SAXS pattern with fiber symmetry has been recorded. In the first case we take the measured intensity $g\left(\varphi\right) \approx I\left(\varphi, s_{hkl} = const\right)$ for the unnormalized pole figure. In the second case we can choose a reflection that is smeared on spherical arcs and project in radial direction over the range of the reflection. From the measured or extracted intensities $I\left(\varphi, s = const\right)$ we then compute the orientation parameter by numerical integration and normalization

$$f_{or}\left(I\left(\varphi, s = const\right)\right) = \frac{\int_0^{\pi/2} I\left(\varphi, s\right) \frac{1}{2}\left(3\cos^2\varphi - 1\right)\sin\varphi\,d\varphi}{\int_0^{\pi/2} I\left(\varphi, s\right)\sin\varphi\,d\varphi}. \qquad (9.10)$$

The uniaxial orientation parameter related to the orientation of polymer chains gains particular importance, because it can also be determined by measurement of birefringence [250, 252].

If the observed reflections are not on spherical arcs, the computation of an orientation parameter becomes an arbitrary operation that is not exclusively related to misorientation of structure. Most probably the topology of the structural entities is coupled to their orientation[6], and Chap. 10 applies.

### 9.1.3 Character of Fiber-Symmetrical Orientation Distributions

**Unimodal** is a fiber-symmetrical orientation distribution $g\left(\varphi\right)$ defined in $\varphi \in [0, \pi/2]$, if it monotonously decays to the right and to the left of a single maximum.

**Meridional** is a function $g\left(\varphi\right)$, if its maximum is at $\varphi = 0$. Such distributions describe fiber orientation.

**Equatorial** is a function $g\left(\varphi\right)$, if its maximum is at $\varphi = \pi/2$. Such distributions describe film orientation.

**Split-Meridional Distribution.** Figure 9.4a displays a case of a unimodal, *non-meridional* orientation distribution: the most probable orientation of the structural entities does not coincide with the meridian. If the orientation distribution itself is broad, the split character of the distribution may be invisible. Then it is an *apparently* meridional distribution.

**Split-Equatorial Distributions** analogously, have their maximum symmetrically split above and beneath the equator.

---

[6]Such coupling may be caused from a deformation of the structural entities that is varying as a function of their orientation with respect to the principal axis of the material.

Thus if the most probable orientation of the structural entities is increasingly tilted away from the equator or out of the meridian, the character of the orientation distribution stays apparently equatorial or meridional, respectively – until a splitting of the maxima becomes detectable (cf. Fig. 9.8).

## 9.2 Observed Intensity and Oriented Intensity – The Relation

After having considered the basics of preferred orientation, we now approach the problem of how to separate orientation from the intrinsic scattering intensity of the structural entities. After some reasoning the general relation for the observed intensity

$$I(s) = \iiint g(\xi) \bar{I}_{opt} (T(\xi)s) \, d\xi_1 d\xi_2 d\xi_3 \tag{9.11}$$

expressed in terms of the orientation distribution, $g(\xi)$, and the "perfectly oriented" scattering intensity, $\bar{I}_{opt}(s)$, of "the average structural entity" is obtained (RULAND [253]). Here $\xi = (\xi_1, \xi_2, \xi_3)$ stands for any suitable parameterization of rotation[7], e.g., the 3 EULER angles or 3 direction cosines. $T(\xi)$ is the rotator – a tensor of rank 2 – represented by a $3 \times 3$ matrix. There are peculiar properties enjoyed by rotators (in particular under Fourier transform) that are utilized to derive Eq. (9.11). Equation (9.11) is only valid if the distances among the entities are sufficiently random and uncorrelated with the rotator. This approximation holds for the WAXS almost always, for the SAXS frequently.

In Eq. (9.11) polydispersity of the structural entities is considered, and the average intensity $\bar{I}_{opt}$ describes the perfectly oriented *representative* structural entity in our material. It is coupled to the other quantity that is of interest – the orientation distribution $g(\varphi, \psi, \varphi')$ – here written in another parameterization of the rotation, with the extra $\varphi'$ denoting a rotation of the representative structural entity about an axis of its own.

## 9.3 Desmearing by Use of a Master Orientation Distribution

In a lucky case we may already have found a smeared reflection, $I_{hkl}$, of which we know that it is extremely sharp in $\bar{I}_{opt}$. This means that

$$I_{hkl}(s) = \delta \left( s - \frac{1}{d_{hkl}} \right) g_{hkl}(\varphi, \psi) \tag{9.12}$$

the intensity of the found reflection is an image of $g(\varphi, \psi)$ placed on a spherical shell with the radius $1/d_{hkl}$ in reciprocal space. We thus may directly use this image to "orientation-desmear" the scattering pattern as a whole. If no such "master

---

[7]EULER's Theorem: *Rotation is the general movement of a rigid body in space with a single point fixed.*

orientation distribution" is found in the pattern, one may obtain a good analytical approximation by application of RULAND's streak method (Sect. 9.7).

In practice, orientation desmearing is carried out by expanding[8] the found pole figure in spherical harmonics (Eq. (9.8)). Then the multipole expansion of the whole scattering pattern is computed[9]. By weighted division of the two sets of expansion coefficients (Eq. (9.17)), the expansion coefficients of the *orientation-desmeared* scattering pattern are obtained. Finally, from these coefficients the desmeared pattern is reconstructed using the multipole expansion equation.

After the reconstruction, a *cross-check* should show that the reference reflection is degenerated to a $\delta$-distribution, and there are *no negative intensities in the desmeared image*. If this is not the case, the found reference peak was broadened not only by imperfect orientation[10]. In this case an iterative trial-and-error method is helpful: the peak is proportionally narrowed, until over-desmearing can no longer be detected. The equations mentioned are directly applicable in case of fiber symmetry. If the symmetry of the scattering pattern is lower, the simplification must be reverted to a set of equations in the complete spherical harmonics instead of the Legendre polynomials.

## 9.4 F2: Double Fiber Symmetry – Simplified Integral Transform

Finding an image of the orientation distribution in the measured scattering pattern is a rare stroke of luck. Nevertheless, fiber symmetry $I(\mathbf{s}) = I(s, \varphi)$ is frequently observed – and even if $I_{opt}(\mathbf{s})$ does not show fiber symmetry, at least the scattering of the *average representative* scattering entity, $\bar{I}_{opt}(\mathbf{s}) = \bar{I}_{opt}(s, \varphi)$, is frequently fiber symmetrical. This is the case, if in the material rotation of the scattering entities about their principal axis is not hindered. Under these premises the material exhibits "double fiber-symmetry" (F2).

RULAND [253] shows that in this case the integral transform Eq. (9.11) can be simplified and solved. The corresponding geometrical relationships are sketched in Fig. 9.5.

> More simple solutions are found for special cases. Already in 1933 KRATKY [248] has presented a method for the case in which the observed orientation distribution has its maximum on the equator. In 1979 the problem treated by KRATKY has been revisited by LEADBETTER and NORRIS [254]. They present a different solution which is frequently applied in studies of liquid-crystalline polymers. BURGER and RULAND [255] pinpoint the error in the deduction of LEADBETTER and

---

[8]Each coefficient of the multipole expansion is computed by a numerical integration – after aligning and normalizing the found orientation distribution.

[9]This function is already aligned with its fiber axis in $s_3$-direction, and we do not normalize it. Together with the normalization of the reference peak these measures guarantee the conservation of the 3D scattering intensity under the desmearing operation.

[10]Think of instrumental broadening, finite lattice size.

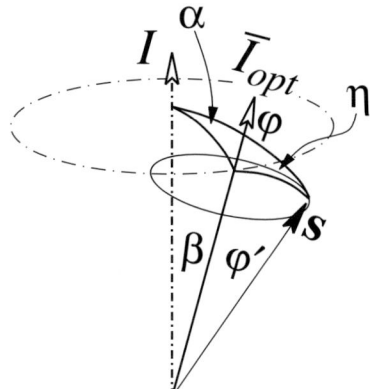

**Figure 9.5.** 3D geometrical relations in the scattering pattern for the case of double fiber symmetry (F2). *Dash-dotted* are both the axis of the observed pattern, $I$, and one of its reflection circles. Drawn in *solid line* are both the axis of a tilted representative structural entity and a reflection circle of its fiber-symmetrical intensity, $\bar{I}_{opt}$. Important for the simplification are the relations in the *spherical triangle* plotted in **bold**

NORRIS. They show that the result of KRATKY is, indeed, a special case ($\varphi' = \pi/2$) of RULAND's general treatment [253, 256], which is sketched in the sequel.

In the deduction RULAND determines, which contribution to the observed intensity, $I$, is added by each reflection *ring* of the likewise fiber symmetrical function, $\bar{I}_{opt}$. Then he adds up all the rings weighted by the orientation function $g(\beta)$. In this way Eq. (9.11) is simplified. A general solution is obtainable by multipole expansion.

The result of the simplified integral equation is

$$I(\varphi) = \int_0^{\pi/2} \bar{I}_{opt}(\varphi') \, F(\varphi, \varphi') \, \sin\varphi' d\varphi' \tag{9.13}$$

with

$$F(\varphi, \varphi'_j) = 2 \int_0^{\pi} g\left(\beta\left(\varphi, \varphi'_j, \eta\right)\right) d\eta \tag{9.14}$$

denoting the intensity on a spherical shell of constant radius. We observe that this kernel of the orientation smearing, $F(\varphi, \varphi')$ is an orientation distribution itself. It is a function both of the angle $\varphi$ of the observed fiber-symmetrical intensity, and of the angle $\varphi'$ measured in the local coordinate system of the perfectly oriented, average structural entity. Relations to the angle $\eta$ are established in the spherical triangle of Fig. 9.5 from the spherical law of cosines.

## 9.5 F3: $g(\varphi)$ Shows Fiber Symmetry – Solution

Of highest practical relevance is the case in which the scattering pattern, the structural entities, and even the orientation distribution $g(\varphi, \psi) = g(\varphi)$ show uniaxial symmetry (F3-materials). If the structure is ruled by polydispersity and the material is uniaxially oriented, F3 is most frequently fulfilled. In this case the mathematical relations are considerably simplified. Suitably the orientation distribution is normalized

$$\iint g(\varphi) \, d\omega = 2\pi \int_0^\pi g(\varphi) \sin\varphi \, d\varphi = 1. \tag{9.15}$$

Its simplified multipole expansion is

$$g(\varphi) = \sum_{\ell=0}^\infty a_\ell P_\ell(\cos\varphi). \tag{9.16}$$

The coefficients of the multipole expansion are computed from Eq. (9.8), and after analogous expansions of both the intensity of the perfectly oriented structural entity ($\bar{I}_{opt}$, $b_\ell$), and of the measured intensity ($I$, $c_\ell$), RULAND [253] obtains a set of algebraic equations among the expansion coefficients,

$$b_\ell = \frac{2\ell + 1}{4\pi} \frac{c_\ell}{a_\ell}, \tag{9.17}$$

which solves the problem exactly and without approximation. In order to arrive at this equation, the integral equation Eq. (9.13), the addition theorems for Legendre polynomials, and the relation $\int_0^\pi \cos(m\eta) \, d\eta = 0$ for $m \neq 0$ are used.

In practice, we have to carry out one numerical integration for the computation of each coefficient. Nevertheless, the coefficients for small index $\ell$ will be most important, and the coefficients with odd index $\ell$ are vanishing anyway from central symmetry. Thus the numerical effort can be mastered.

Finally, we still have a problem: $g(\varphi)$ is not yet known except for the simple case mentioned on p. 197 in Eq. (9.12).

## 9.6 Extraction of $g(\varphi)$ from Meridional or Equatorial Reflections

### 9.6.1 Unimodal Meridional Reflection Intensity

If the smeared image of a point-reflection on the meridian is unimodal, the orientation distribution $g(\varphi)$ is readily accessible. In this case a "texture measurement" through the reflection, $I(\varphi, s = c)$ contains the information sought-after [256]

$$I(\varphi, s = c) = \lim_{\varphi \to 0} F(\varphi, \varphi') = 2\pi g(\varphi). \tag{9.18}$$

In Fig. 9.5 two sides of the spherical triangle merge ( $\varphi' = \varphi$), which leads to the result. Thus the shape of the meridional reflection at constant $|s|$ directly reflects the shape of the orientation distribution.

### 9.6.2 Unimodal Equatorial Reflection Intensity

For a unimodal equatorial reflection the treatment is more involved. If the distribution is narrow, it follows from Fig. 9.5 an approximation $\varphi' \approx \pi/2$, which can be used to obtain an approximative solution, in turn (RULAND cited by THÜNEMANN [257], p. 28)

$$F(\varphi, \varphi') = 2 \int_0^{\frac{\pi}{2}} \Re \left( K_{eq}\left(\beta, \frac{\pi}{2} - \varphi + \varphi'\right) + K_{eq}\left(\beta, \frac{\pi}{2} - \varphi - \varphi'\right) \right) g(\beta) \sin\beta \, d\beta,$$

(9.19)

with

$$K_{eq}(\beta, \varphi) = \frac{1}{\left(\sin^2 \varphi - \cos^2 \beta\right)^{1/2}}.$$

(9.20)

On each side of the meridian, only a *distorted* image of the orientation distribution is observed. Nevertheless, even equatorial reflections can generally be used for the purpose of orientation-desmearing if we make assumptions concerning the analytical type of the orientation distribution. The corresponding method is demonstrated in the following section.

## 9.7 The Ruland Streak Method

**Motivation and Principle.**   Broadened reflections are characteristic for soft matter. The reason for such broadening is predominantly both the short range of order among the particles in the structural entities, and imperfect orientation of the entities themselves. A powerful method for the separation of these two contributions is RULAND's streak method [30–34]. Short range of order makes that the reflection is considerably extended in the radial direction of reciprocal space – often it develops the shape of a streak. This makes it practically possible to measure reflection breadths separately on *several*[11] nested shells in reciprocal space. As a function of shell diameter one of the contributions is constant, whereas the other is changing[12]. If the measurement is performed on spheres (azimuthal), the orientation component is constant.

**History.**   WILKE [129] considers the case that different orders of a reflection are observed and that the orientation distribution can be analytically described by a Gaussian on the orientation sphere. He shows how the apparent increase of the integral breadth with the order of the reflection can be used to separate misorientation effects from size effects. RULAND [30–34] generalizes this concept. He considers various analytical orientation distribution functions [9, 84, 124] and deduces that the method can be used if only a *single* reflection is sufficiently extended in radial direction, as is frequently the case with the streak-shaped reflections of the anisotropic

---

[11] Thus we study pole figures *as a function of* $|\mathbf{s}|$.

[12] The same principle is the fundament of the WARREN-AVERBACH method (cf. Sect. 8.2.5.5) for the separation of size and distortion of structural entities. Thus the mathematics is partially identical.

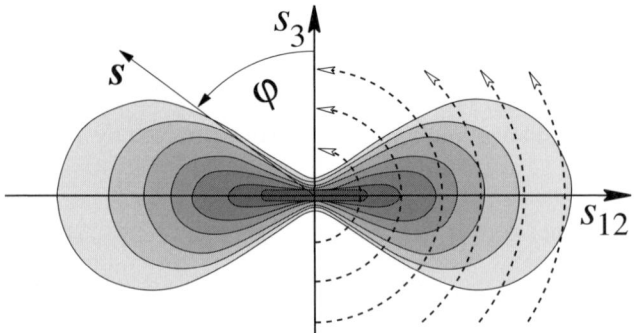

**Figure 9.6.** Fanning-out of an equatorial streak in a fiber pattern caused from misorientation. Dashed arcs indicate azimuthal scans that are performed in practical measurements. The recorded scattering curves are used to separate the effects of misorientation and extension of the structural entities

structural entities[13] observed in fiber materials. In recent publications [34, 258, 259] several authors demonstrate, both how this concept is flexibly modified, and how it helps to characterize the perfection of nanostructured materials.

**Basic Ideas.** Let the average structural entity be anisotropic with fiber symmetry. Let its shape

$$Y(\mathbf{r}) = Y_{12}(r_{12}) Y_3(r_3) \tag{9.21}$$

be factorized in cylindrical coordinates[14]. As the structural entity is perfectly oriented, even its scattering intensity

$$\bar{I}_{opt}(\mathbf{s}) \propto \Phi_{12}^2(s_{12}) \Phi_3^2(s_3) \tag{9.22}$$

is factorized in cylindrical coordinates.

Let us consider a frequent problem: the scattering of elongated voids or of microfibrils is investigated. In such materials an equatorial streak is observed – similar to the one sketched in Fig. 9.6. If the voids were perfectly oriented, its integral breadth measured as a function of $s_{12}$

$$B_{obs}(s_{12}) = \int_{-\infty}^{\infty} I(s_{12}, s_3) \, ds_3 / I(s_{12}, 0) \tag{9.23}$$

would be

$$B_{obs}(s_{12}) = \frac{1}{\langle L \rangle} \tag{9.24}$$

---

[13] elongated voids, rods, microfibrils, layers, layer stacks

[14] In the practice of materials science this is rarely a restriction, because any higher complexity of its outer shape is generally smeared from polydispersity.

constant. $\langle L \rangle$ is the average longitudinal extension[15] [30, 31, 34] of the voids, rods or microfibrils. If misorientation has to be taken into account, the shape of the streak is fanned out according to the orientation distribution $g(\varphi)$. Concerning the form factor $\Phi_3^2(s_3)$ of the rigid structural entities it follows in good approximation [30]

$$I(s, \varphi) \simeq \frac{1}{2}\Phi_{12}^2(s) \left[ g(\varphi) *_\psi \Phi_3^2(s\varphi) \right],$$
(9.25)

that the orientation distribution becomes broadened by an *angular convolution*, $*_\varphi$, of an image of the shape factor $\Phi_3^2(s\varphi)$ that continuously narrows (argument $s\varphi$), as the radius of the arc in Fig. 9.6, $s$, is increased.

Separation of the two components is accomplished by means of data, in which the apparent azimuthal integral breadth

$$B_{obs}(s) = \frac{1}{I(s, \pi/2)} \int_{-\pi/2}^{\pi/2} I(s, \varphi) \, d\varphi$$
(9.26)

is determined as a function of the arc radius in reciprocal space (cf. Fig. 9.6).

The decrease of the apparent breadth $B_{obs}(s)$ with increasing $s$ is a function of the analytical shape of the orientation distribution $g(\varphi)$.

If a *Gaussian* can be used to describe the orientation distribution it follows

$$B_{obs}^2(s) = \frac{B_p^2}{s^2} + \frac{1}{s^2 \langle L \rangle^2} + B_g^2.$$

Here $B_p$ describes the inevitable instrumental broadening by the known integral breadth of the primary beam[16], and $B_g$ is the true integral breadth of the orientation distribution. For the determination of $\langle L \rangle$ and $B_g$ the relation is linearized

$$s^2 B_{obs}^2(s) = B_p^2 + \frac{1}{\langle L \rangle^2} + s^2 B_g^2.$$
(9.27)

If a *Lorentzian* is a proper model for the orientation distribution, instead of Eq. (9.27) the relation [31, 32, 34]

$$s B_{obs}(s) = B_p + \frac{1}{\langle L \rangle} + s B_g$$
(9.28)

is obtained. In practice, both variants are tested for the best linearization using the data of the experiment. Thereafter the structural parameters are determined from the best plot. An example is shown in Fig. 9.7.

---

[15] For worm-shaped entities we will not determine the extension, but only the persistence length of the entities (cf. p. 165).

[16] We need $B_p$ in the *perpendicular* direction with respect to the direction in which the streak is extending. The beam profile has (hopefully) been measured after the adjustment of the instrument and before installation of the beam stop. Before being used in Eq. (9.27) or Eq. (9.28) $B_p$ is converted to reciprocal space units.

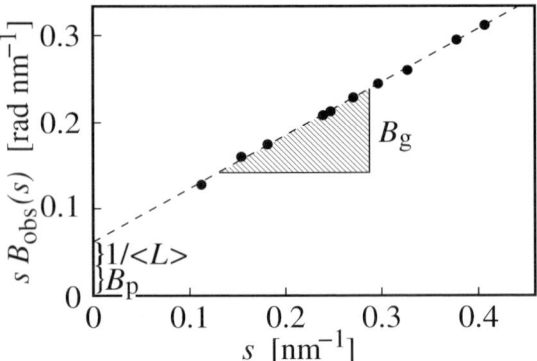

**Figure 9.7.** Separation of misorientation ($B_g$) and extension of the structural entities ($1/\langle L \rangle$) for known breadth of the primary beam ($B_p$) according to RULAND's streak method. The perfect linearization of the observed azimuthal integral breadth measured as a function of arc radius, $s$, shows that the orientation distribution is approximated by a Lorentzian with an azimuthal breadth $B_g$

After analysis by means of the streak method, a good analytical approximation for the orientation distribution is known. Orientation desmearing becomes possible. For this purpose the method described in Sect. 9.5 can be utilized.

If meridional streaks are found for materials built from layer stacks, these patterns can be analyzed analogously [259]. An application to data sets combined from series of reflections with increasing order is possible, as well.

If the most-probable orientation of the structural entities is no longer parallel to the fiber axis, we may observe a clearly inclined streak. Such orientations are frequently observed in herbal and animal natural fibers [45, 260, 261]. If the split nature of the orientation distribution is clearly detected, the streak method can be applied or adapted.

Problems arise, as the orientation distribution starts to split, but the split nature is not yet discernible. THÜNEMANN [257] is discussing this problem in his thesis. He describes, how to determine the true tilt angle of the structural entities, and he determines the minimum tilt angle that is required for the split nature to become detectable (Fig. 9.8). We observe that, in practice, a split nature of Lorentzian orientation distributions (solid line) is detected earlier than a split nature of Gaussians – at least up to an apparent[17] integral breadth of 70°. The reason is that Lorentzians are more pointed than Gaussians – in the vicinity of their maximum.

---

[17] *Why is the determined integral breadth only an apparent one?* Lorentzians show a high background – and this background accumulates in particular, when Lorentzians are used as orientation distributions and thus are wrapped around the sphere. Thus the minimum of a Lorentzian orientation distribution on a sphere is not zero and the integral breadth that is determined by subtracting the minimum is only an apparent one.

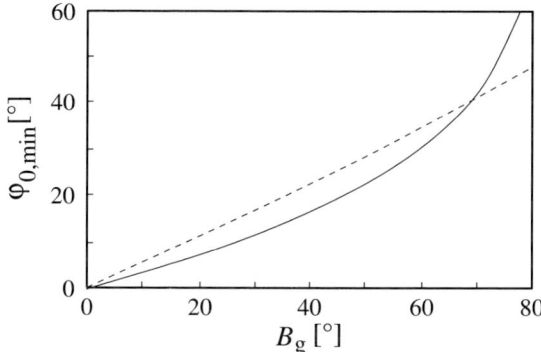

**Figure 9.8.** Minimum average tilt angle, $\varphi_{0,min}$, of structural entities measured with respect to the fiber axis at which the *split nature of the orientation distribution* becomes observable – plotted as a function of the integral breadth $B_g$ of the orientation distribution $g(\varphi)$. *Solid line:* $g(\varphi)$ is a Lorentzian. *Dashed:* Gaussian

## 9.8 Analytical Functions Wrapped Around Spheres: Shape Change

**The Fitting Problem.**   In many studies in particular of natural fibers, orientation distributions are *picked from* spherical arcs in scattering patterns and then fitted by Gaussians or Lorentzians. The result is the finding of an isotropic background. At least part of this background is not related to structure, but to a fundamental misunderstanding.

The reason is the following: as we use analytical functions to model orientation, we take a function with, in general, *infinite* support and map it on the orientation sphere. Because of the *inherent periodicity of the spherical surface*, the branches of the analytical function wrap around the sphere up to infinity (Fig. 9.9). As the orientation distribution is, again, *picked from* the scattering pattern, it does not unwrap, again. This wrapping must be considered. RULAND [256] shows that the periodic superposition of shifted distribution images may not only result in an intrinsic background. Moreover, every shape is fundamentally changed. So if we put a Gaussian on the sphere, we receive $\cos^{2\nu}\varphi$ in return. Lorentzians become Poisson kernels – with considerable intrinsic background.

**Pathway to the Solution.**   All orientation functions are defined on the orientation sphere. At the best, their period is $\pi$. Thus the mapping of an analytical function $h(x)$ on the orientation sphere is equivalent to

$$g(\varphi) = \sum_{m=-\infty}^{\infty} h(\varphi - m\pi), \qquad (9.29)$$

and $g(\varphi)$ is observed. For $h(x)$ RULAND [33, 256, 262] is discussing Gaussians, Lorentzians and even a superposition of the two kinds of distributions [262].

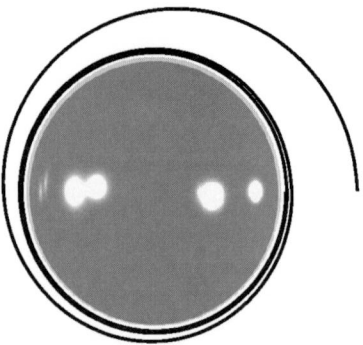

**Figure 9.9.** Infinite functions in a periodic world. Using a function (*black line*) as an orientation function means to wrap it around the orientation sphere. Only one branch of a LORENTZ distribution at the right side of the equator is sketched. Shape change occurs

**Solution for GAUSS Distributions.** For a Gaussian $h(x) = \exp\left(-v(\varphi - n\pi)^2\right)$ Eq. (9.29) becomes

$$g_G(\varphi) = \cos^{2v}(\varphi) = \pi \sum_{n=-\infty}^{\infty} \exp\left(-v(\varphi - n\pi)^2\right) \tag{9.30}$$

with the integral breadth

$$B_G = \int \exp\left(-v\varphi^2\right) d\varphi = \sqrt{\pi/v}. \tag{9.31}$$

The equations are analytical for even and positive values of $v$.

**Solution for LORENTZ Distributions.** For LORENTZ distributions the solution is the Poisson kernel

$$g_L(\varphi) = \frac{1-q^2}{1+q^2-2q\cos(2\varphi)} = \sum_{n=-\infty}^{\infty} \frac{2\log(1/q)}{(\log(1/q))^2 + 4(\varphi - m\pi)^2} \tag{9.32}$$

with its integral breadth

$$B_L = \frac{\pi}{2} \log\left(\frac{1}{q}\right) \tag{9.33}$$

for $0 < |q| < 1$. $q$ is the breadth parameter, and the Poisson kernel enjoys several interesting properties. For $q = 0$, the distribution degenerates into the *isotropic distribution*. For negative values of $q$, the distribution is a unimodal equatorial distribution. For positive $q$, the maximum is on the meridian. For $|q| = 1$, the orientation distribution becomes infinitesimally narrow. Because of the fact that LORENTZ distributions are slowly decaying, their mapping on the orientation sphere always results in an intrinsic background. Thus $B_L$ is not accessible from measured data. In practice, the zero level will be put in the observed minimum of the orientation distribution $g(\varphi)$ on the orientation sphere, and an apparent integral breadth

$$B_{La} = \frac{\pi}{2}(1 - q) \tag{9.34}$$

is determined. This result is obtained from Eq. (9.33) by taking limits $q \to 1$.

# 10 Orientation Growing from the Isotropic State

Straining of isotropic materials is a common method of testing or processing. During such treatment uniaxial orientation is frequently growing (Fig. 10.1).

**Rigid Structural Entities.**   If the initial structure is described by rigid, anisotropic structural entities which are oriented at random, the evolution of anisotropic scattering is readily studied by means of the methods presented in Chap. 9. A practical example is the study of growing orientation in fiber-reinforced materials.

**Shape Change of Structural Entities.**   In many cases the growing anisotropy is not only a phenomenon of rotating structural entities, but also goes along with a *deformation of the structural entities* themselves. This case will be studied here. Only affine deformations shall be discussed. In practice, such processes are observed while thermoplastic elastomers are subjected to mechanical load, but also while fibers are spun.

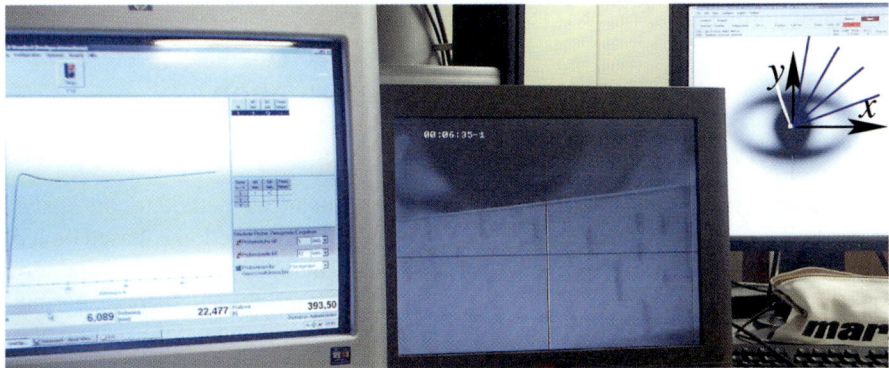

**Figure 10.1.** USAXS observation during straining of an SBS block copolymer. *Right monitor:* Intensity maxima on an ellipse. Raw-data coordinate system $(x,y)$ and radial cuts for data analysis are indicated. *Middle:* Videotaping of sample. *Left:* Stress-strain curve. Control booth of beamline BW4, HASYLAB, Hamburg

**The General Experimental Observation.**   Only the analysis of the most frequent experimental observation [248, 263–266] is discussed. In corresponding studies scattering curves are extracted in radial direction, and the positions of the found
reflection maxima are plotted *vs.* the polar angle $\psi'$ between the slicing direction and
the fiber axis. As long as the material is isotropic, the reflection maxima are found on
a spherical shell, the Debye sphere. As orientation is growing, the maxima are found
on ellipsoidal shells, the ellipticity of which is increasing with increasing *draw ratio*,
$\lambda_d = \varepsilon + 1$.

**Evaluation Methods.**   Two evaluation methods are discussed in this chapter. RU-
LAND's theory is based on reasoning concerning structure in 3D space. The strict
approach relates structure to the scattering pattern. For example, the microscopic
draw ratio of the structural entities can be determined. The other approach is the
MGZ-technique of scattering image processing. Based on 2D master-peaks mapped
by an elliptical transformation, almost any SAXS fiber pattern can readily be modeled. Long periods or tilt angles determined by the method are not more reasonably
founded in scattering theory as are direct determinations from peak position and peak
angle. Nevertheless, because the whole pattern is considered in the fit, small changes
can be determined with higher significance. Thus the method is valuable for compar-
ative studies. Moreover, it can be used to properly align raw patterns or to reconstruct
patterns that are distorted – e.g., from accidental tilt of the specimen in the sample
holder.

   If the intended evaluation can be carried out on isotropic material, and thus the
observed anisotropy is rather an obstacle than an advantage, the fiber pattern can be
isotropized (cf. Sect. 8.4.2). This may, in particular, be helpful if lamellar structures
are analyzed. If the focus of the study is on the anisotropic structure, the multidimensional CDF (cf. Sect. 8.5.5) may be a suitable tool for analysis. Several studies
have demonstrated the power of the CDF method for the study of structure evolution
during straining [174, 177, 181–183].

## 10.1 R          's Theory of Affine Deformation

### 10.1.1 Overview

In a fundamental paper [265] RULAND develops an advanced method for the analysis
of scattering patterns showing moderate anisotropy. The deduction is based on a 3D
model and the concept of highly oriented lattices. The addition of distortion terms
makes sure that the theory is applicable to distorted structures and their scattering.

   According to his deduction the common finding of ellipsoidal deformation of the
reflections is indicative for affine deformation. Moreover, he arrives at an equation
that permits to determine with high accuracy the microscopical draw ratio, $\lambda_d$, of
the structural entities from the ellipticity of the deformed Debye sphere. This value
can be compared to the macroscopical draw ratio. Even the intensity distribution
along the ellipsoidal ridge is predicted for a bcc-lattice of spheres, and deviations of
experimental data are discussed.

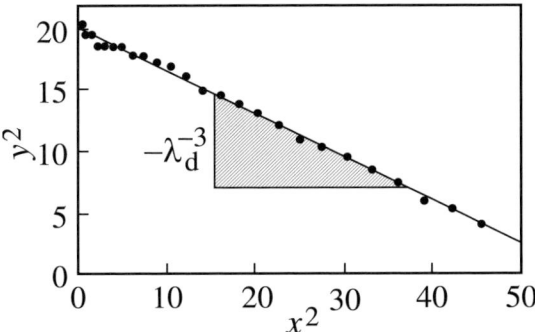

**Figure 10.2.** Plot of the positions $(x, y)$ of peak maxima extracted on radial rays in a moderately oriented SAXS fiber pattern according to BRANDT & RULAND [265]. A microscopical draw ratio $\lambda_d = 1.41$ of the structural entities is determined from the slope

## 10.1.2 Application

Let $(s_{12}, s_3)$ be the coordinates of reflection maxima determined on radial rays in the scattering pattern. Then a linearizing plot of the ellipsoidal shape is

$$s_3^2 = b^2 - \frac{b^2}{a^2} s_{12}^2, \tag{10.1}$$

with $a$ and $b$ being the semimajor and the semiminor axis of the rotation-ellipsoid, respectively. Then the equation [265]

$$\frac{b^2}{a^2} = \frac{1}{\lambda_d^3} \tag{10.2}$$

relates the slope in the linearizing plot from Eq. (10.1) to the microscopical draw ratio of the structural entities. The intercept

$$b^2 = \frac{1}{L_m^2} \tag{10.3}$$

in the linearizing plot is related to the extrapolated long period, $L_m$, in meridional direction (fiber axis). Figure 10.2 demonstrates the linearizing plot according to Eq. (10.1) for positions $(x, y)$ of peak maxima extracted along radial rays from the 2D SAXS scattering pattern. $x$ and $y$ are in length units on the image (cf. Fig. 10.1). If $x = c s_{12}$ and $y = c s_3$ are valid, the draw ratio can be determined directly, without transformation from length units to the units of reciprocal space.

The clear linearity of the data demonstrates the affine character of the deformation. From the intercept, $b^2$, the semiminor axis of the rotation ellipsoid is determined. After transformation to units of reciprocal-space, the meridional long period follows from Eq. (10.3).

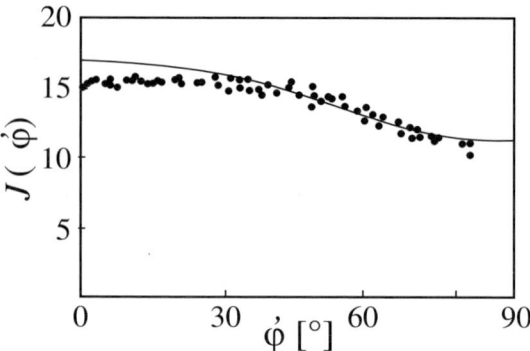

**Figure 10.3.** Shape of the maximum peak intensity, $J(\varphi')$, extracted from radial sections of a moderately anisotropic ($\lambda_d = 1.41$) SAXS pattern with fiber symmetry as a function of the sectioning angle $\varphi'$ related to the fiber axis. *Dots:* experimental values. *Solid line:* Theoretical shape according to RULAND [265]

Modeling the initial structure by spherical domains in a bcc-lattice[1], the theoretical *intensity along the ellipsoidal ridge* as a function of the angle $\psi'$ between fiber axis and the direction of the radial beam is

$$J_h(\varphi') \propto g(\lambda_d, \varphi') \, |\langle\Phi\rangle_D|^2 (s_h') \, H_z^2(s_h'),\tag{10.4}$$

with

$$g(\lambda_d, \varphi') = \frac{1}{\lambda_d} \sqrt{\frac{1 + \lambda_d^{-3}\tan^2\varphi'}{1 + \lambda_d^{-6}\tan^2\varphi'}}.\tag{10.5}$$

and

$$s_h' = s_h \sqrt{\frac{1}{\lambda_d}\sin^2\varphi' + \lambda_d^2\cos^2\varphi'}.$$

$|\langle\Phi\rangle_D|^2 (s_h') \, H_z^2(s_h')$ is the *form factor envelope*[2] of the scattering, made from the form factor of ideal spheres and the attenuation term describing a smooth transition of the density at the phase boundary (cf. p. 124).

The comparison between experimental and predicted intensity (Fig. 10.3) reveals deviations between theory and experimental data already at relatively low uniaxial deformation. A reasonable explanation is an increase of lattice distortions by local tensions which modify the envelope and thus the total intensity. Theoretical computation of mechanical anisotropy in a bcc-lattice supports the explanation [265].

---

[1]bcc: body-centered cubic arrangement of the spheres

[2]The form factor is, in fact, an envelope, because it limits the visibility of the reflections: Outside the region where it has decayed to virtually 0, no scattering is observed. Sometimes the envelope is visible: We "see" the spherical, the cylindrical or the layer shape of the fundamental domains in the oriented material.

## 10.2 The MGZ Technique of Elliptical Coordinates

**Applicability.**   Just like RULAND in the previous section, MURTHY, GRUBB and ZERO [266–269], start from the common observation (Fig. 10.1). They propose a simple and powerful parameterization of scattering images of strained materials. Nevertheless, the method does not consider the 3D character (reflections on rotation ellipsoids, 3D structure model): movement of a peak from the meridian to the equator will decrease its observed "image-intensity", although such a decrease is – to a first approximation – a result of distributing the same intensity on a ring in space with a bigger diameter. Anyway, the intensity issue is not the focus of the MGZ method. Compared to the peak shape issue it is a more complex problem, as RULAND has shown (cf. p. 212).

**Elliptical Coordinates.**   Scattering patterns can be expressed as a function of various coordinates. For isotropic scattering patterns, $I(\mathbf{s}) = I(s, \varphi, \psi)$, it is reasonable to choose polar coordinates, $(s, \varphi, \psi)$, because the intensity is factorized $I(\mathbf{s}) = I(s)\, I_\varphi(\varphi)\, I_\psi(\psi)$ in these coordinates and two of the factors $I_\varphi(\varphi)\, I_\psi(\psi) = 1$ are constant.

Transferred to the observation that the reflections in moderately anisotropic scattering images are found on ellipses[3], it appears reasonable to parameterize such images in elliptical coordinates $(u, v)$. The transformation relations are [266]

$$s_{12} = \sqrt{A^2 + u^2}\cos v \tag{10.6}$$

$$s_3 = u \sin v. \tag{10.7}$$

Here $A$ is the distance of the foci, which are found on the $s_{12}$-axis. For $u = 0$ we have plane polar coordinates. Varying $v \in [0, 2\pi]$ at constant $u$ describes an elliptical orbit with $a = \sqrt{A^2 + u^2}$ and $b = u$ its semimajor and semiminor axis, respectively.

If the elliptical parameterization shall be used for automatic alignment of scattering patterns, control of ellipse rotation and displacement are important. Rotation is controlled by subtraction of a constant from $v$. A displaced center is readily considered by subtraction on the left side of Eqs. (10.6) and (10.7).

Transferring RULAND's relation between the ellipticity and the microscopical draw ratio, $\lambda_d$, to the parameterization of MURTHY, GRUBB and ZERO, we receive

$$\frac{A^2}{u^2} = \lambda_d^3 - 1. \tag{10.8}$$

Figure 10.4 shows the first steps of scattering data analysis by means of of the MGZ technique. Replacing $\tan\phi = s_{12}/s_3$ in Eq. (10.1)

$$L_\phi^2 = L_M^2 + L_E^2 \tan^2\phi \tag{10.9}$$

is obtained. The equation describes the variation of the long period as a function of the polar angle, $\phi$. The relations to the geometrical positions of the peak maximum in

---

[3]By not addressing scattering *patterns* and rotation-*ellipsoids* we stress the 2D character of the method.

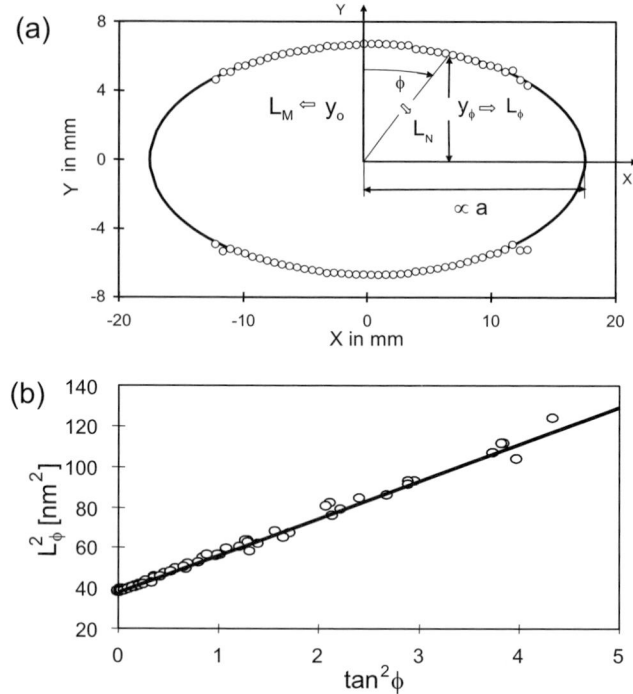

**Figure 10.4.** MGZ analysis of an undrawn polyamide 6 fiber. (a) Plot of the peak maximum position as a function of the detector coordinates $(x, y)$. Relations to the parameters of the ellipse and the structure (long periods $L_M$, $L_\phi$) are indicated. (b) Separation of meridional long period, $L_M$, from the long period $L_\phi$. (courtesy S. MURTHY)

the scattering pattern are $L_\phi = 1/s_3$, $L_E = 1/a$, and $L_M = 1/b$. Here $s_3$ is not a variable, but the meridional component of the peak maximum expressed in units of the reciprocal space. The peak maxima are found on an ellipse with the semimajor axis $a$ and the semiminor axis $b$. The validity of Eq. (10.9) to $\phi$ almost 90° is demonstrated in Fig. 10.4b using data obtained from an undrawn polyamide 6 fiber [269].

**Parameterization of Reflections in Elliptical Coordinates.**     The authors of the technique define the intensity

$$I_j(u, v) = C_j f_j\left(u, u_{0,j}, u_{w,j}\right) g_j\left(v, v_{o,j}, v_{w,j}\right)$$

of the $j$-th model reflection by the product of two 1D functions, $f_j$ and $g_j$, and a scaling factor $C_j$. $u_{0,j}$ and $v_{0,j}$ define the center of the reflection. $u_{w,j}$ and $v_{w,j}$ determine the breadths. A suitable class of functions for the 1D distributions is empirically found by study of the intensity curves in radial sections (cf. Fig 10.1) of the images. According to MURTHY, GRUBB and ZERO [266] a suitable class are Pearson-VII functions with shape factor $m = 2$.

Fiber symmetry makes that every function $g_j$ is the sum of four quadrant functions,

$$g_j\left(v, v_{0,j}, v_{w,j}\right) = \frac{1}{4}\left(h_j\left(v, v_{0,j}, v_{w,j}\right) + h_j\left(v, \pi - v_{0,j}, v_{w,j}\right) + \\ h_j\left(v, -v_{0,j}, v_{w,j}\right) + h_j\left(v, \pi + v_{0,j}, v_{w,j}\right)\right). \qquad (10.10)$$

Thus every reflection in the fiber diagram is defined by one function $f_j\left(u, u_{0,j}, u_{w,j}\right)$ and four quadrant functions. If the model shall be fitted to a scattering pattern in which the fiber is tilted with respect to the primary beam, weighting factors are attached to each of the quadrant functions. After the fit of an experimental scattering image, the found factors quantify the tilt, and the corresponding distortion of the scattering pattern can be eliminated.

# 11 Fitting Models to Data

Because of the importance of model fitting in the field of scattering, some hints are presented and sources for further reading are given.

Models are fitted to scattering data by means of nonlinear regression [270] and related computer programs [154, 271]. The quality of the parameterization (by structural parameters) and of the fit are estimated. The "best fitting model" is accepted. The found values for the structural parameters are plotted *vs.* environmental parameters of the experiment and discussed. Environmental parameters that come into question are, for example, the materials composition, its temperature, elongation, or the elapsed time.

## 11.1 Which Data Are Fitted?

Almost any transformation of data is changing the weight of inherent features that we want to know about. A striking example is the simple logarithmic representation of scattering data and the related distortion of the error-bar spread (p. 124, Fig. 8.11). As we are interested in structure, we should fit data that present an undistorted view of structure. Our X-ray instrument has already transformed structure information into a scattering pattern, and we have to ask what we should do with the pattern before fitting – leave it as it is or transform it back?

**In the Scattering Pattern: The Perfection.** Because infinite lattices (and Gaussians) are the only functions that do not change their shape under Fourier transform, there is a simple answer to the question for the crystallographer: leave the diffraction pattern as it is. It is a true image of structure. The same is valid for scattering data of identical particles (Sect. 8.6) or for nanocomposites with high perfection (Sect. 8.8). In all these cases it is reasonable to implement all necessary corrections into a model function of the scattering intensity, and then to directly[1] fit the recorded data.

The problem starts as soon as the structure is governed by polydispersity. In this case the transformation carried out by the X-ray instrument is no longer true, and

---

[1] Theoretical considerations have been carried out by BURGER and RULAND [20] for the case of ideal structure. According to their deduction it turns out that the discontinuities found in the CLDs, IDFs, or CDFs of ensembles arranged from ideal and identical geometrical bodies seriously aggravate any pre-evaluation or transformation of measured scattering curves.

the scattering intensity presents a strongly biased view of structure information (cf. Sect. 2.6).

**In the Chord Distribution Pattern: The Diversity.** In Sect. 2.6 we have demonstrated that the back-transformation into a CLD, IDF or CDF corrects the biased view on polydispersity that is inherent to scattering data. So at least for the purpose of structure *visualization* one should carry out the back-transformation in order to view polydispersity and complexity of the structure without biased weighting. Nevertheless, this does not answer the question concerning model fitting: *is it good practice to fit models to IDFs that originate from scattering data after pre-evaluation, smoothing and Fourier transformation?* Theoretical considerations for polydisperse soft matter have not yet been carried out, but practical experience indicates that the structure description obtained by fitting a model to the raw scattering data of a polydisperse structure is, in general, not in agreement with the computed IDF of the material. Moreover, the convergence of regression algorithms using polydisperse model functions is rather poor when they are applied to raw scattering data which do not show narrow peaks. On the other hand, fits on the IDF converge quite fast and the transformation of the structural data back to the scattering curve fits quite well [125].

## 11.2 Which Techniques Are Applied?

**Model Fitting.** In general, the residuals between the model function $M(\mathbf{p},x)$ and $m$ pairs of variates are summed

$$R(\mathbf{p}) = \sum_{i=1}^{m} (y_i - M(\mathbf{p},x_i))^2,$$

yielding the function that is minimized[2]. It is called the "residual sum of squares" (RSS). Minimization is carried out by variation of the parameter set $\mathbf{p}$. $\mathbf{p}$ is a vector with $\ell$ components. $\ell$ is the number of model parameters. $R(\mathbf{p})$ is a function defined in *parameter space*.

In the ideal case the program finds the absolute minimum of RSS in parameter space. With good starting values for the (structure) parameters this is possible. In practice, the regression program frequently stops in a local minimum – i.e., in a trough at a higher level than the absolute minimum. Thus model fitting becomes an issue of iteration and trial-and-error. It starts from an iterative optimization of the starting values for the regression program[3], and continues with the optimization

---

[2] In a minimum the gradient $\nabla R(\mathbf{p}) = 0$ is vanishing.

[3] After each run the found parameters will be checked for the necessity to move them away from extreme values (0, big values), and restart the program repeatedly until no further reduction of the RSS is achieved.

of an initial model[4], the comparison of different[5] good models, and ends with the acceptance of the best model fit. If only one data set shall be fitted, the significance of the outcome is difficult to assess. If a whole series of scattering data is fitted as a function of external parameters, the workload is high, but the significance can be assessed by comparison and by plausibility considerations.

**Regression Algorithms.** The fitting of structural models to X-ray scattering data requires utilization of *nonlinear* regression techniques. The respective methods and their application are exhausted by DRAPER and SMITH [270]. Moreover, the treatment of nonlinear regression in the *"Numerical Recipes"* [154] is recommended.

Two extremely different regression techniques are the *gradient methods*[6] and the *simplex algorithm*[7]. The already mentioned validation methods are all based on the gradient. Nevertheless, the simplex method [271] is the better scrutinizer in parameter space. Thus it appears reasonable to combine [197] both methods: after an iterative regression run of the simplex, the ultimate gradient is computed and the quality of the fit is validated. For the purpose of gradient computation, it is no longer necessary to analytically compute the partial derivatives of the model function with respect to the parameters, $\partial M(\mathbf{p})/\partial p_i$, $i \in [1, \ell]$ ($\ell$ parameters). The replacement of the differential by the difference

$$\frac{\partial M(\mathbf{p}, x)}{\partial p_i} \approx \frac{M(\mathbf{p}, x) - (\mathbf{p}'_i, x)}{\varepsilon}$$

with $\mathbf{p}'_i = (p_1, p_2, \ldots p_{i-1}, p_i + \varepsilon, p_{i+1}, \ldots)$ is sufficient, if the central parts of the algorithm are using variables with 80 bit accuracy. The increment $\varepsilon$ can be derived from the accuracy request of the scientist who is using the program. Now validation can be started. From the components of the ultimate gradient, first the *matrix of the normal equations* [270]

$$Z_{j,k}(\mathbf{p}) = \left[ \sum_{i=1}^{\ell} \frac{\partial M(\mathbf{p}, x_i)}{\partial p_j} \frac{\partial M(\mathbf{p}, x_i)}{\partial p_k} \right] \tag{11.1}$$

is computed. Its inverse is $Z_{j,k}^{-1}$ (from now on we omit the argument for clarity). $F = m + 1 - \ell$ is the number of *degrees of freedom*, and $R(\mathbf{p})$ is the ultimate RSS after the regression run. Then

---

[4]The model must neither have too many nor too few parameters. Are there too few parameters, it cannot fit the data. Are there too many parameters, some of them are correlated and wide error bars are attached to them.

[5]In general, several models should be tested. Systematic deviations between model and data indicate, how the model should be varied in order to improve the fit. If anisotropic materials have been studied, the CDF frequently exhibits the ingredients of a good model.

[6]After each iterative improvement all the gradient methods compute the gradient $\nabla R(\mathbf{p})$. The way they use the gradient is different for the variants (steepest descent, Marquardt-method,...) of the gradient method.

[7]The state of the simplex is not stored in 1 parameter set, but in a polygon with $\ell + 1$ parameter sets in $\ell$-dimensional parameter space. The polygon can move, grow and shrink in parameter space. It tries to encircle the minimum sought-after.

$$c_i = \sqrt{\left| Z_{i,i}^{-1} \right| R(\mathbf{p}) / F}$$

is the asymptotic interval of confidence for the $i$-th parameter value. The asymptotic *parameter-correlation matrix*, $C_{i,j}$, is defined by

$$C_{i,j} = Z_{i,j}^{-1} / \sqrt{Z_{i,i}^{-1} Z_{j,j}^{-1}},$$

and demonstrates the correlation among the model parameters. Thus a report of the regression program should not only return the parameter values, but also the corresponding error bars. Overparameterization is detected in the correlation matrix. A value $\left| C_{i,j} \right| > 0.95$ indicates overparameterization. In this case one of the parameters $i$ or $j$ is superfluous. Underparameterization is detected in a plot that shows the deviations of the measured points from the model. These deviations should look random. Unfortunately this is not frequently the case when models are fitted to IDFs. From RSS, number of measured points $m$, and the number of model parameters, $\ell$, a model-free validation parameter

$$E = \sqrt{\frac{R(\mathbf{p})}{m(m-\ell)}}$$

is computed, the so-called *"estimated error of the fit"*, EEF. On the basis of EEF the best fitting model can be selected. As a whole the method gains significance if large series of data are analyzed. In this case outliers which have stopped in a local minimum are easily detected. Frequently their fits can be improved by taking better starting values from the results of their neighbors in the series.

**Conditioning Data.**    The efficiency – and sometimes even the success – of regression and smoothing procedures is a function of the data presented to the program. One should not offer curves to regression or smoothing algorithms running on digital computers, in which the ranges of the variates differ too much. If the $r$ values range between 0 and 200, the values of the IDF $g_1(r)$ should be mapped to a similar range. The optimum conditioning procedure is to map both the $x$-values and the $y$-values of the data to the interval $[0, 0.5]$ by an affine transformation. Such mapping is described by a displacement and a scaling factor. It is easily inverted upon return from the algorithm.

The effectivity of data conditioning is explained as follows. In regression algorithms and smoothing procedures, constructs like the one presented in Eq. (11.1) have to be evaluated by means of a digital computer. Extensive computation of differences, their multiplication and summing occurs. This means that rounding errors may accumulate. In this case either the convergence behavior becomes erratic or slow. Sometimes the algorithm even stops, because a matrix is "not well-conditioned", i.e., it contains both very big and very small numbers.

# References

[1] Debye P, Menke H (1931) Erg techn Röntgenkunde 2:1
[2] Stribeck N (2006) J Appl Cryst 39:237
[3] Fedorova IS, Schmidt PW (1978) J Appl Cryst 11:405
[4] Abramowitz M, Stegun IA (eds.) (1968) Handbook of Mathematical Functions. Dover Publications, New York
[5] Hosemann R, Bagchi SN (1962) Direct Analysis of Diffraction by Matter. North-Holland, Amsterdam
[6] Guinier A (1963) X-Ray Diffraction. Freeman, San Francisco
[7] Alexander LE (1979) X-Ray Diffraction Methods in Polymer Science. Wiley, New York
[8] Ruland W (1964) Br J Appl Phys 15:1301
[9] Ruland W, Smarsly B (2002) J Appl Cryst 35:624
[10] Elsner G, Riekel C, Zachmann HG (1984) In: Kausch HH, Zachmann HG (eds.), *Advances in Polymer Science*, vol. 67, pp. 1–58, Springer, Berlin
[11] Kirfel A, Eichhorn K (1990) Acta Cryst A46:271
[12] Kahn R, Fourme R, Gadet A, Janin J, Dumas C, André D (1982) J Appl Cryst 15:330
[13] Macgillavry CH, Rieck GD (eds.) (1968) International Tables for X-ray Crystallography, vol. 2. Physical and Chemical Tables. The Kynoch Press, Birmingham
[14] Ruland W (1961) Acta Cryst 14:1180
[15] Weisstein EW (1999), Fourier Transform. From MathWorld - A Wolfram Web Resource. http://mathworld.wolfram.com/FourierTransform.html
[16] Bonart R (1966) Kolloid Z u Z Polymere 211:14
[17] Debye P, Bueche AM (1949) J Appl Phys 20:518
[18] Porod G (1951) Colloid Polym Sci 124:83
[19] Brychkov YA, Glaeske HJ, Prudnikov AP, Tuan VK (1992) Multidimensional integral transformations. Gordon & Breach, Philadelphia
[20] Burger C, Ruland W (2001) Acta Cryst A57:482
[21] Schmidt PW, Brill L. O (1967) In: Rowell RR, Stein RS (eds.), *Electromagnetic Scattering. Proceedings of ICES 2, Amherst, MA, June 1965*, pp. 169–186, Gordon & Breach, New York
[22] Baltá Calleja FJ, Vonk CG (1989) X-Ray Scattering of Synthetic Polymers. Elsevier, Amsterdam
[23] Conner WC, Webb SW, Spanne P, Jones KW (1990) Macromolecules 23:4742
[24] Schroer CG, Kuhlmann M, Roth SV, Gehrke R, Stribeck N, Almendarez Camarillo A, Lengeler B (2006) Appl Phys Lett 88:164102
[25] Spontak RJ, Williams MC, Agard DA (1988) Polymer 29:387
[26] Stribeck N (2001) J Appl Cryst 34:496
[27] Stokes AR (1948) Proc Phys Soc 61:382
[28] Warren BE, Averbach BL (1950) J Appl Phys 21:595
[29] Warren BE, Averbach BL (1952) J Appl Phys 23:497
[30] Ruland W (1969) J Polym Sci Part C 28:143

[31] Perret R, Ruland W (1969) J Appl Cryst 2:209

[32] Ruland W (1968) J Appl Cryst 1:90

[33] Perret R, Ruland W (1970) J Appl Cryst 3:525

[34] Thünemann AF, Ruland W (2000) Macromolecules 33:1848

[35] Fraser RD, Macrae TP, Miller A, Rowlands RJ (1976) J Appl Cryst 9:81

[36] Buhmann MD (2000) Acta Numerica 9:1

[37] VNI, pv-wave manuals. V 7.5 (2001), Boulder, Colorado

[38] Rudolf PR, Landes BG (1994) Spectroscopy Eugene Oreg 9:22

[39] Hammersley     AP,     FIT2D     V12.012     Reference     Manual.
     http://www.esrf.fr/computing/scientific/FIT2D/

[40] RSI, Interactive Data Language IDL. V 6.1 (2004), Boulder, Colorado

[41] Rasband   W,   ImageJ  -  Image   processing   and   analysis   in   Java.
     http://rsb.info.nih.gov/ij/

[42] Haberäcker P (1989) Digitale Bildverarbeitung. Hanser, München

[43] Stribeck N, Buchner S (1997) J Appl Cryst 30:722

[44] Riekel C, Engström P (1995) Nuclear Instr Meth Phys Res B97:224

[45] Müller M, Czihak C, Vogl G, Fratzl P, Schober H, Riekel C (1998) Macro-
     molecules 31:3953

[46] Waigh TA, Donald AM, Heidelbach F, Riekel C, Gidley MJ (1999) Biopoly-
     mers 49:91

[47] Assouline E, Wachtel E, Grigull S, Lustiger A, Wagner HD, Marom G (2001)
     Polymer 42:6231

[48] Dreher S, Zachmann HG, Riekel C, Engstrøm P (1995) Macromolecules
     28:7071

[49] Heidelbach F, Riekel C, Wenk HR (1999) J Appl Cryst 32:841

[50] Wang YD, Cakmak M (2001) Polymer 42:4233

[51] Loidl D, Paris O, Burghammer M, Riekel C, Peterlik H (2005) Phys Rev Lett
     95:225501

[52] Stribeck N, Almendarez Camarillo A, Nöchel U, Schroer C, Kuhlmann M,
     Roth SV, Gehrke R, Bayer RK (2006) Macromol Chem Phys 207:1239

[53] Bark M, Zachmann HG (1993) Acta Polym 44:259

[54] Wutz C, Bark M, Cronauer J, Döhrmann R, Zachmann H (1995) Rev Sci
     Instrum 66:1303

[55] Kolb R, Seifert S, Stribeck N, Zachmann HG (2000) Polymer 41:1497

[56] Stribeck N, Almendarez Camarillo A, Cunis S, Bayer RK, Gehrke R (2004)
     Macromol Chem Phys 205:1445

[57] Stribeck N, Bayer R, Bösecke P, Almendarez Camarillo A (2005) Polymer
     46:2579

[58] Casselyn M, Finet S, Tardieu A, Delacroix H (2002) Acta Cryst D58:1568

[59] Rössle M, Panine P, Urban VS, Riekel C (2004) Biopolymers 74:316

[60] Weiss TM, Narayanan T, Wolf C, Gradzielski M, Panine P, Finet S, Helsby
     WI (2005) Phys Rev Lett 94:038303

[61] Burger HC, van Cittert PH (1932) Z Phys 79:722

[62] Heikens D (1959) J Polym Sci 35:139

[63] Porod G (1961) Fortschr Hochpolym Forsch 2:363

[64] DuMond JWM (1947) Phys Rev 72:83
[65] Guinier A, Fournet G (1955) Small-Angle Scattering of X-Rays. Chapman and Hall, London
[66] Ruland W (1977) Colloid Polym Sci 255:417
[67] Ruland W (1978) Colloid Polym Sci 256:932
[68] Glatter O (1974) J Appl Cryst 7:147
[69] Wiegand W, Ruland W (1979) Colloid Polym Sci 257:449
[70] Stribeck N (1996) Macromolecules 29:7217
[71] Schelten J, Hendricks RW (1975) J Appl Cryst 8:421
[72] Pedersen JS (2004) J Appl Cryst 37:369
[73] Schuster M, Göbel H (1995) J Phys D Appl Phys 28:A270
[74] Schuster M, Göbel H (1996) J Phys D Appl Phys 29:1677
[75] Müller-Buschbaum P, Roth SV, Burghammer M, Diethert A, Panagiotou P, Riekel C (2003) Europhys Lett 61:639
[76] Lengeler B, Tümmler J, Snigirev A, Snigireva I, Raven C (1998) J Appl Phys 84:5855
[77] Schroer CG, Kuhlmann M, Lengeler B, Günzler TF, Kurapova O, Benner B, Rau C, Simionovici AS, Snigirev AA, Snigireva I (2002) Proc SPIE 4783:10
[78] Hendrix J (1984) In: Kausch HH, Zachmann HG (eds.), *Advances in Polymer Science*, vol. 67, pp. 59–98, Springer, Berlin
[79] Wilson KS (1998) Nature Struct Biol 5:627
[80] Stribeck N (2003) Anal Bioanal Chem 376:608
[81] Luzzati V (1957) Acta Cryst 10:643
[82] Perret R, Ruland W (1971) J Appl Cryst 4:444
[83] Ruland W, Tompa H (1972) J Appl Cryst 5:1
[84] Ruland W, Smarsly B (2004) J Appl Cryst 37:575
[85] Klug HP, Alexander LE (1974) X-Ray Diffraction Procedures for Polycrystalline and Amorphous Materials. 2nd edn., John Wiley & Sons, New York
[86] Feigin LA, Svergun DI (1987) Structure Analysis by Small-Angle X-Ray and Neutron Scattering. Plenum Press, New York
[87] Kratky O, Porod G, Kahovec L (1951) Z Elektrochemie 55:53
[88] Perret R, Ruland W (1972) J Appl Cryst 5:116
[89] Polizzi S, Stribeck N, Zachmann HG, Bordeianu R (1989) Colloid Polym Sci 267:281
[90] Bösecke P, Diat O (1997) J Appl Cryst 30:867
[91] Orthaber D, Bergmann A, Glatter O (2000) J Appl Cryst 33:218
[92] Stribeck N, Ruland W (1978) J Appl Cryst 11:535
[93] Wendorff JH, Fischer EW (1973) Colloid Polym Sci 251:876
[94] Rathje J, Ruland W (1976) Colloid Polym Sci 254:358
[95] Wiegand W, Ruland W (1979) Progr Colloid Polym Sci 66:355
[96] Hermans PH, Heikens D, Weidinger A (1959) J Polym Sci 35:145
[97] Warren BE (1990) X-Ray Diffraction. Dover, New York
[98] Blake FC (1933) Rev Mod Phys 5:169
[99] Girard E, Legrand P, Roudenko O, Roussier L, Gourhant P, Gibelin J, Dalle D, Ounsy M, Thompson AW, Svensson O, Cordier MO, Robin S, Quiniou R,

Steyer JP (2006) Acta Cryst D62:12

[100] Glocker R (1971) Materialprüfung mit Röntgenstrahlen. 5th edn., Springer, Berlin

[101] Glatter O, Kratky O (eds.) (1982) Small Angle X-ray Scattering. Academic Press, London

[102] Macgillavry CH, Rieck GD (eds.) (1968) International Tables for X-Ray Crystallography, vol. III. Kynoch Press, Birmingham

[103] Wang ZG, Hsiao BS, Sirota EB, Srinivas S (2000) Polymer 41:8825

[104] Bras W, Dolbnya I, Detollenaere D, van Tol R, Malfois M, Greaves G, Ryan A, Heeley E (2003) J Appl Cryst 36:791

[105] Liu J, Geil PH (1997) J Macromol Sci Phys B36:61

[106] Stribeck N, Wutz C (2001) J Polym Sci Part B Polym Phys 39:1749

[107] Brandrup J, Immergut EH, Grulke EA, Abe A, Bloch DR (eds.) (1999) Polymer Handbook. 4th edn., John Wiley & Sons, New York

[108] Vonk CG (1973) J Appl Cryst 6:148

[109] Prosa TJ, Moulton J, Heeger AJ, Winokur MJ (1999) Macromolecules 32:4000

[110] Krenzer E, Ruland W (1981) Colloid Polym Sci 259:405

[111] Wcislak L, Klein H, Bunge HJ, Garbe U, Tschentscher T, Schneider JR (2002) J Appl Cryst 35:82

[112] Dehlinger U, Kochendörfer A (1939) Z Kristallograf 101:134

[113] Kochendörfer A (1939) Z Kristallograf 101:149

[114] Kochendörfer A (1944) Z Kristallograf 105:393

[115] Kochendörfer A (1944) Z Kristallograf 105:438

[116] Zernike F, Prins JA (1927) Z Phys 41:184

[117] Hosemann R (1962) Polymer 3:349

[118] Tchoubar D, Méring J (1969) J Appl Cryst 2:128

[119] Blöchl G, Bonart R (1986) Makromol Chem 187:1525

[120] van Berkum JGM, Vermeulen AC, Delhez R, de Keijser TH, Mittemeijer EJ (1994) J Appl Cryst 27:345

[121] Buchanan DR, Miller RL (1966) J Appl Phys 37:4003

[122] Warren BE (1941) J Appl Phys 12:375

[123] Hall WH (1949) Proc Phys Soc A62:741

[124] Ruland W (1965) Acta Cryst 18:581

[125] Stribeck N (1993) Colloid Polym Sci 271:1007

[126] Stribeck N (1993) J Phys IV 3:507

[127] Bonart R, Hosemann R, McCullough RL (1963) Polymer 4:199

[128] Hermans JJ (1944) Rec Trav Chim Pays Bas 63:211

[129] Hosemann R, Wilke W (1964) Faserforsch Textiltechnik 15:521

[130] Santa Cruz C, Stribeck N, Zachmann HG, Baltá Calleja FJ (1991) Macromolecules 24:5980

[131] Perret R, Ruland W (1971) Colloid Polym Sci 247:835

[132] Ruland W (1971) J Appl Cryst 4:70

[133] Ruland W (1975) Progr Colloid Polym Sci 57:192

[134] Vonk CG (1973) J Appl Cryst 6:81

[135] Koberstein JT, Morra B, Stein RS (1980) J Appl Cryst 13:34

[136] Higgins JS, Benoît HC (1994) Polymers and Neutron Scattering. Clarendon Press, Oxford

[137] Porod G (1952) Colloid Polym Sci 125:51

[138] Jánosi A (1983) Monatsh f Chemie 114:377

[139] Stribeck N (2000) ACS Symp Ser 739:41

[140] Ruland W (1987) Macromolecules 20:87

[141] Méring J, Tchoubar D (1968) J Appl Cryst 1:153

[142] Wolff T, Burger C, Ruland W (1994) Macromolecules 27:3301

[143] Pfeifer P, Ehrburger-Dolle F, Rieker TP, T. GM, Hoffman WP, Molina-Sabio M, Rodríguez-Reinoso F, Schmidt PW, Voss DJ (2002) Phys Rev Lett 88:11502

[144] Schmidt PW (1991) J Appl Cryst 24:414

[145] Rudin SA (1999) In: Weber Robert L.; Mendoza E (ed.), *A Random Walk in Science*, pp. 98–99, Inst. of Physics, London

[146] Avnir D, Biham O, Lidar D, Malcai O (1998) Science 279:39

[147] Ruland W (2001) Carbon 39:323

[148] Desper CR, Stein RS (1967) J Polym Sci Part B Polym Lett 5:893

[149] Jánosi A (1986) Z Phys B 63:383

[150] Blundell DJ (1978) Polymer 19:1258

[151] Barnes JD, Kolb R, Barnes K, Nakatani AI, Hammouda B (2000) J Appl Cryst 33:758

[152] Barnes JD, Bras W (2003) J Appl Cryst 36:664

[153] Stribeck N (2002) Colloid Polym Sci 280:254

[154] Press WH, Teukolsky SA, Vetterling WT, Flannery BP (1992) Numerical Recipes. Cambridge University Press, Cambridge

[155] Flores A, Pietkiewicz D, Stribeck N, Roslaniec Z, Baltá Calleja FJ (2001) Macromolecules 34:8094

[156] Claver Jr. GC, Buchdahl R, Miller RL (1956) J Polym Sci 20:202

[157] Peterlin A (1972) Text Res J 42:20

[158] Hosemann R (1949) Z Phys 127:16

[159] Vonk CG, Kortleve G (1967) Colloid Polym Sci 220:19

[160] Kortleve G, Vonk CG (1968) Colloid Polym Sci 225:124

[161] Duhamel P, Hollman H (1984) Electronics Letters 20:14

[162] Goderis B, Reynaers H, Koch MHI, Mathot VBF (1999) J Polym Sci Part B Polym Phys 37:1715

[163] Strobl GR, Schneider M (1980) J Polym Sci Part B Polym Phys B18:1343

[164] Vonk CG, Pijpers AP (1985) J Polym Sci Part B Polym Phys 23:2517

[165] Crist B (2000) J Macromol Sci Phys B39:493

[166] Crist B (2001) J Polym Sci Part B Polym Phys 39:2454

[167] Crist B (2003) Macromolecules 36:4880

[168] Vonk CG (1979) Colloid Polym Sci 257:1021

[169] Méring J, Tchoubar-Vallat D (1965) C R Acad Sc Paris 261:3096

[170] Méring J, Tchoubar-Vallat D (1966) C R Acad Sc Paris 262:1703

[171] Stribeck N (1992) Colloid Polym Sci 270:9

[172] Stribeck N (1980) Computation of the Lamellar Nanostructure of Polymers by Computation and Analysis of the Interface Distribution Function from the Small-Angle X-ray Scattering. Ph.D. thesis, Phys. Chem. Dept., University of Marburg, Germany

[173] Stribeck N, Fakirov S, Sapoundjieva D (1999) Macromolecules 32:3368

[174] Stribeck N, Fakirov S (2001) Macromolecules 34:7758

[175] Fronk W, Wilke W (1985) Colloid Polym Sci 263:97

[176] Wilke W, Bratrich M (1991) J Appl Cryst 24:645

[177] Stribeck N, Buzdugan E, Ghioca P, Serban S, Gehrke R (2002) Macromol Chem Phys 203:636

[178] Stribeck N, Bayer R, von Krosigk G, Gehrke R (2002) Polymer 43:3779

[179] Barbi V, Funari SS, Gehrke R, Scharnagl N, Stribeck N (2003) Macromolecules 38:749

[180] Barbi V, Funari SS, Gehrke R, Scharnagl N, Stribeck N (2003) Polymer 44:4853

[181] Stribeck N, Androsch R, Funari SS (2003) Macromol Chem Phys 204:1202

[182] Stribeck N, Fakirov S, Apostolov AA, Denchev Z, Gehrke R (2003) Macromol Chem Phys 204:1000

[183] Stribeck N, Funari SS (2003) J Polym Sci Part B Polym Phys 41:1947

[184] Stribeck N (2004) Macromol Chem Phys 205:1455

[185] Stribeck N, Almendarez Camarillo A, Bayer R (2004) Macromol Chem Phys 205:1463

[186] Stribeck N, Bösecke P, Bayer R, Almendarez Camarillo A (2005) Progr Coll Polym Sci 130:127

[187] Brumberger H (ed.) (1995) Modern Aspects of Small-Angle Scattering, vol. 451 of *NATO ASI Series, Series C*. Kluwer, Dordrecht

[188] Svergun DI, Koch MHJ (2003) Rep prog phys 66:1735

[189] Vonk CG (1976) J Appl Cryst 9:433

[190] Glatter O (1977) J Appl Cryst 10:415

[191] Glatter O (1979) J Appl Cryst 12:166

[192] Letcher JH, Schmidt PW (1966) J Appl Phys 37:649

[193] Debye P (1915) Ann Phys 46:809

[194] Petoukhov MV, Svergun DI (2003) J Appl Cryst 36:540

[195] Svergun DI, Petoukhov MV, Koch MHJ (2001) Biophys J 80:2946

[196] Konarev PV, Volkov VV, Sokolova AV, Koch MHJ, Svergun DI (2003) J Appl Cryst 36:1277

[197] Stribeck N (1989) Colloid Polym Sci 267:301

[198] Statton WO (1962) J Polym Sci 58:205

[199] Statton WO (1968) Z Kristallogr 127:229

[200] Stribeck N (1999) J Polym Sci Part B Polym Phys 37:975

[201] Cohen Y, Thomas EL (1987) J Polym Sci Part B Polym Phys B25:1607

[202] Titchmarsh EC (1948) Introduction to the Theory of Fourier Integrals. Clarendon Press, Oxford

[203] Marichev OI (1983) Handbook of Integral Transforms of Higher Transcendental Functions. Ellis Horwood Ltd., Chichester

[204] Schmidt PW (1967) J Math Phys 8:475
[205] Wolfram-Research (2005) Mathematica. Version 5.2. Wolfram Research, Inc., Champaign, Illinois
[206] Porod G (1972) Monatsh Chem 103:395
[207] Förster S, Burger C (1998) Macromolecules 31:879
[208] Kilian HG, Wenig W (1974) J Macromol Sci Phys B9:463
[209] Schultz JM, Lin JS, Hendricks RW (1978) J Appl Cryst 11:551
[210] Schultz JM, Fischer EW, Schaumburg O, Zachmann HG (1980) J Polym Sci Polym Phys 18:239
[211] Hosemann R (1950) Kolloid Z 117:13
[212] Kinning DJ, Thomas EL (1984) Macromolecules 17:1712
[213] Santos A, Yuste SB, López de Haro M (2002) J Chem Phys 117:5785
[214] Kamiyama T, Sasaki M, Suzuki K (2000) J Appl Cryst 33:447
[215] Kamiyama T, Suzuki K (1998) J Non cryst solids 232-234:476
[216] Boyer D, Tarjus G, Viot P (1995) J Chem Phys 103:1607
[217] Percus JK, Yevick GJ (1958) Phys Rev 110:1
[218] Bonnier B, Boyer D, Viot P (1994) J Phys A 27:3671
[219] Rényi A (1963) Sel Transl Math Stat Prob 4:203
[220] Rényi A (1958) Publ Math Inst Budapest 3:109
[221] Evans JW (1993) Rev Mod Phys 65:1281
[222] Kumar P (2002) J Inequal Pure and Appl Math 3:art. 41
[223] Balakrishnan N, Gupta SS (1998) In: Balakrishnan N.; Rao CR (ed.), *Handbook of Statistics*, vol. 17, chap. 2, pp. 25–59, Elsevier, Amsterdam
[224] Aggarwala R, Balakrishnan N (1996) Ann Inst Statist Math 48:757
[225] Varghese P, Braswell R, Wang B, Zhang C (1996) Comput Aided Des 28:723
[226] Burgos E, Bonadeo H (1987) J Phys A 20:1193
[227] Brämer R, Ruland W (1976) Makromol Chem 177:3601
[228] Weick D, Hosemann R (1980) Colloid Polym Sci 258:593
[229] Brämer R (1972) Colloid Polym Sci 250:1034
[230] Strobl GR (1973) J Appl Cryst 6:365
[231] Gelfer M, Burger C, Fadeev A, Sics I, Chu B, Hsiao BS, Heintz A, Kojo K, Hsu SL, Si M, Rafailovich M (2004) Langmuir 20:3746
[232] Vineyard GH (1982) Phys Rev B Condens Matter 26:4146
[233] Sinha SK, Sirota EB, Garoff S, Stanley HB (1988) Phys Rev B Condens Matter 38:2297
[234] Pynn R (1992) Phys Rev B Condens Matter 45:602
[235] Rauscher M, Salditt T, Spohn H (1995) Phys Rev B 52:16855
[236] Lazzari R (2002) J Appl Cryst 35:406
[237] Lee B, Park I, Yoon J, Park S, Kim J, Kim KW, Chang T, Ree M (2005) Macromolecules 38:4311
[238] Lee B, Park YH, Hwang YT, Oh W, Yoon J, Ree M (2005) Nat Mater 4:147
[239] Lee B, Yoon J, Oh W, Hwang Y, Heo K, Jin KS, Kim J, Kim KW, Ree M (2005) Macromolecules 38:3395
[240] Smarsly B, Gibaud A, Ruland W, Sturmayr D, Brinker CJ (2005) Langmuir 21:3858

[241] Ruland W, Smarsly B (2005) J Appl Cryst 38:78
[242] Benedetti A, Polizzi S, Riello P, Pinna F, Goerigk G (1997) J Catalysis 171:345
[243] Goerigk G, Haubold HG, Lyon O, Simon JP (2003) J Appl Cryst 36:425
[244] Friedel G (1913) C R Acad Sci 157:1533
[245] Lyon O, Guillon I, Servant B (2001) J Appl Cryst 34:484
[246] Ballauff M (2001) Curr Opin Colloid Interface Sci 6:132
[247] Fratzl P (2003) J Appl Cryst 36:397
[248] Kratky O (1933) Kolloid Z 64:213
[249] Kratky O (1933) Kolloid Z 68:347
[250] Hermans PH, Platzek P (1939) Kolloid Z 88:68
[251] Ward IM (ed.) (1997) Structure and Properties of Oriented Polymers. Chapman and Hall, London
[252] Pepper RE, Samuels RJ (1988) In: Mark HF, et al. (eds.), *Encyclopedia of Polymer Science and Engineering*, vol. 14, 2nd edn., pp. 261–298, Wiley, New York
[253] Ruland W (1977) Colloid Polym Sci 255:833
[254] Leadbetter AJ, Norris EK (1979) Mol Phys 38:669
[255] Burger C, Ruland W (2006) J Appl Cryst 39:889
[256] Ruland W, Tompa H (1968) Acta Cryst A24:93
[257] Thünemann A (1995) Structural transitions and microphase formation in homopolymers, copolymers and metallic carbonyles of acrylonitrile. Ph.D. thesis, Universität Marburg
[258] Thünemann AF, Ruland W (2000) Macromolecules 33:2626
[259] Keum JK, Burger C, Hsiao BS, Somani R, Yang L, Chu B, Kolb R, Chen H, Lue CT (2005) Progr Colloid Polym Sci 130:113
[260] Reiterer A, Lichtenegger H, Fratzl P, Stanzl-Tschegg SE (2001) J Mater Sci 36:4681
[261] Putthanarat S, Stribeck N, Fossey SA, Eby RK, Adams WW (2000) Polymer 41:7735
[262] Baltá Calleja FJ, Kilian HG (1988) Colloid Polym Sci 266:29
[263] Séguéla R, Prud'homme J (1988) Macromolecules 21:635
[264] Young P, Stein RS, Kyu T, Lin JS (1990) J Polym Sci Part B Polym Phys B28:1791
[265] Brandt M, Ruland W (1996) Acta Polym 47:498
[266] Murthy NS, Zero K, Grubb DT (1997) Polymer 38:1021
[267] Murthy NS, Grubb DT, Zero K (2000) ACS Symp Ser 739:24
[268] Murthy NS, T. GD (2006) J Polym Sci Part B Polym Phys 44:1277
[269] Murthy NS, Grubb DT, Zero K (2000) Macromolecules 33:1012
[270] Draper NR, Smith H (1980) Applied Regression Analysis, Second Edition. John Wiley & Sons, New York
[271] Caceci MS, Cacheris WP (1984) Byte 1984:340

# Subject Index

absolute intensity, 40, 79, 86–94, 96
  calibration
    colloidal suspensions, 134
    glassy carbon, 91
    liquids, 92, 119, 134
    Lupolen standard, 91
    noble metals, 92, 134
    synthetic polymers, 91
    SAXS, 87–92, 134
    WAXS, **92**
absolute measurements, *see* absolute intensity
absorption, **77**
absorption factor
  experimental determiation, 79
absorption law, 77
additive
  clarifying, 165
  nucleating, 165
anisotropy, 26
anomalous dispersion, **190**
approximation
  tangent plane, **12**, 27
ASAXS, **188**
asymmetrical peaks, **101**
autocorrelation, **16**
autocorrelation triangle, 144
AWAXS, **188**
AXRD, **188**
azimuthal averaging, 26, 125, **129**, 130

Babinet's theorem, 132, 179
  resolve ambiguity, 179
background, **69**, **99**
  fluctuation, **118**, 119

  metals, 119
  polymers, 119
  simple fluids, 119
bandlimited function, **25**, 162
beads, 163
beam
  incident, 11
  scattered, 11
beam stop, 37, 100
Bessel functions, **167**
binning, **58**
birefringence, 196
Born
  approximation, 185
  distorted wave approximation, 185
Bragg spacing, **101**, **111**
Bragg's law, 13, 24
breadth
  FWHM, 106
  integral, **24**, 106
    azimuthal, **203**
Breit-Wigner distribution, *see* Lorentz distribution

calibration, *see* absolute intensity, calibration
Cauchy distribution, *see* Lorentz distribution
CCD, 12
CDF, 16, **153**
  multidimensional, **152**
  of uncorrelated particles, **155**
  peaks
    sign, **156**
central projection theorem, *see* theorem, Fourier slice